Theory of NMR Parameters

Theory of NMR Parameters

I. Ando
Department of Polymer Chemistry
Tokyo Institute of Technology
Ookayama, Meguro Ku, Tokyo, Japan

and

G. A. Webb
Department of Chemistry
The University of Surrey
Guildford, Surrey, UK

1983

Academic Press

A Subsidiary of Harcourt Brace Jovanovich, Publishers
London   New York
Paris   San Diego   San Francisco   São Paulo
Sydney   Tokyo   Toronto

ACADEMIC PRESS INC. (LONDON) LTD.
24/28 Oval Road
London NW1 7DX

United States Edition published by
ACADEMIC PRESS INC.
111 Fifth Avenue
New York, New York 10003

Copyright © 1983 by ACADEMIC PRESS INC. (LONDON) LTD.

All rights Reserved
No part of this book may be reproduced in any form by photostat, microfilm, or any other means, without written permission from the publishers

British Library Cataloguing in Publication Data
Ando, I.
  Theory of NMR parameters.
  1. Nuclear magnetic resonance
  I. Title   II. Webb, G.A.
  538'.362   QC762

ISBN 0-12-056820-9

LCCCN 83-70977

Typeset by Eta Services (Typesetters) Ltd, Beccles, Suffolk
and printed in Great Britain by
St Edmondsbury Press, Bury St Edmunds, Suffolk.

Preface

In writing this book we have attempted to provide an introduction to a molecular understanding of the factors which determine the commonly encountered NMR parameters. Relaxation phenomena are normally discussed in the language of quasi-classical mechanics. It has been our aim to restrict ourselves mainly within the confines of quantum chemistry, thus relaxation effects are only dealt with in a cursory manner together with an introduction to nuclear shielding and spin–spin coupling in Chapter 1.

It is assumed that the reader is familiar with the basic concepts of quantum mechanics, consequently Chapter 2 consists of a review of the elementary techniques of quantum chemistry together with introductory accounts of LCAO–MO and VB theories. These are applied to various methods of calculating nuclear shielding and spin–spin coupling in Chapters 3 and 4 respectively.

Chapter 5 consists of listings of SOS and FPT programs for calculating nuclear shielding at the semi-empirical level. These programs have been developed in our own laboratories and are not widely available; in contrast to this comparable programs for spin–spin couplings may be obtained from the Quantum Chemistry Program Exchange, Bloomington, Indiana, USA.

There are at least three requirements of a satisfactory theory of nuclear shielding and spin–spin coupling. First, one would like the theory to provide results which are quantitatively accurate, in which case it could be used for predictive purposes. A second requirement might be that the results, obtained from a given theoretical model, be able to give some chemical detail of the factors which cause nuclear shielding and spin–spin coupling, and perhaps more importantly, shielding and coupling variations amongst members of a series of molecules. Finally, it should be possible to apply a chosen model to molecules of moderate size without requiring excessive amounts of computational effort. We have endeavoured to address some of the problems arising from these requirements in this book.

We wish gratefully to acknowledge many stimulating discussions with our students and the forbearance of our wives who became NMR widows during the preparation of this book.

Tokyo and Guildford I. Ando
August 1983 G. A. Webb

Contents

Preface v

Symbols and Abbreviations ix

1 Introduction to NMR Parameters

1.A	General remarks on NMR	1
	1.A.1 Resonance condition	2
	1.A.2 Experimental aspects	3
	1.A.3 Spectral analysis	3
1.B	Nuclear shielding	4
1.C	Nuclear spin–spin coupling	6
1.D	Nuclear spin relaxation	7
	1.D.1 Relaxation by magnetic dipole interactions	10
	1.D.2 Relaxation by electric quadrupole interactions	13
	1.D.3 Relaxation by spin–rotation interactions	15
	1.D.4 Relaxation by nuclear shielding anisotropy	16
	1.D.5 Relaxation by scalar coupling	16
References		18

2 Introduction to Quantum Theory

2.A	Formalism of quantum chemistry	21
	2.A.1 Schrödinger's equation	21
	2.A.2 The variation method	22
	2.A.3 Perturbation methods	24
2.B	Molecular orbital theory	28
	2.B.1 Wavefunction Ψ	28
	2.B.2 Energy corresponding to Ψ	29
	2.B.3 LCAO self-consistent field (SCF) equations	30
	2.B.4 Unrestricted SCF method	31
	2.B.5 Empirical MO methods	32
	2.B.6 Semi-empirical MO methods	33
	2.B.7 Non-empirical methods	36
2.C	Valence bond method	39
2.D	Applications of quantum chemistry to second-order molecular properties	40
	2.D.1 Nuclear shielding	41
	2.D.2 Nuclear spin–spin couplings	43
References		45

3 Nuclear Shielding

3.A	General theoretical background	47
3.B	Calculations based upon SCF theories	53
3.C	Calculations giving gauge-independent shielding data	57
	3.C.1 FPT calculations	58
	3.C.2 SOS calculations	63
3.D	Ring currents	69
3.E	Medium effects on nuclear shielding	76
3.F	Conclusions	79
References		81

4 Spin–Spin Couplings

4.A	General theoretical background	83
4.B	MO approaches	88
	4.B.1 SOS method	88
	4.B.2 FPT method	92
	4.B.3 SCP method	94
4.C	VB approaches	96
4.D	Calculations of spin–spin couplings	99
	4.D.1 SOS calculations	99
	4.D.2 FPT calculations	100
	4.D.3 SCP calculations	102
	4.D.4 Comparison of the results from different theories and MO treatments	104
	4.D.5 Correlations of spin–spin couplings with other molecular properties	104
4.E	Medium effects on spin–spin couplings	108
References		112

5 Program Listings

5.A	A computer program for calculating nuclear shielding by means of FPT-INDO and CNDO/2 methods	115
5.B	A program for calculating nuclear shielding by means of the INDO/S-SOS procedure	159
References		212
Index		213

Symbols and Abbreviations

These lists contain the symbols and abbreviations most frequently used in this monograph, but they are not intended to be exhaustive. Trivial usage of subscripts, etc., is not always mentioned in the symbols listed below. The symbols for physical constants, e.g. e, h or π, are not included since they follow completely accepted usage.

SYMBOLS

A	vector potential due to applied magnetic field
B	magnetic induction field (magnetic flux density)
B_0	static magnetic field of an NMR spectrometer
B_L	local magnetic field
B_1	r.f. magnetic field associated with v_1
C_x	spin–rotation coupling constant for nucleus x
$C_\parallel, C_\perp$	components of C parallel and perpendicular to molecular symmetry axis
E_n	eigenvalue of $\mathcal{H}$
F	Fock operator
F^n	Slater–Condon repulsion integrals
G^n	Slater–Condon repulsion integrals
H_{ij}	matrix element of $\mathcal{H}$
$\mathcal{H}$	Hamiltonian operator – subscripts indicate its nature
$\mathcal{H}^n$	operator describing a perturbation
I_n	nuclear spin operator for nucleus n
I_{nx}, I_{ny}, I_{nz}	components of I_n
I	ring current
^nJ	nuclear spin–spin coupling interaction through n bonds; further information may be given in brackets or by subscripts
J^c	contact contribution to J
J^o	orbital contribution to J
J^d	dipolar contribution to J
$J(\omega)$	power spectral density
J_{ij}	Coulomb integral in MO theory
K_{ij}	exchange integral in MO theory
$K(r_i)$	geometric factor in ring current theory

SYMBOLS AND ABBREVIATIONS

m_I	magnetic quantum number for nuclear spin I
M	molecular moment of inertia
N	Avogadro's number
p	electronic momentum
p_A	valence p orbital of atom A
$P_{\mu\nu}$	element of bond order, charge density matrix
Q_{AB}	used in AEE theory to denote $P_{\mu\nu}$
Q	Coulomb integral in VB theory
eQ	nuclear quadrupole moment
eq	electric field gradient
$Q^{\alpha\beta}$	imaginary component of bond order, charge density matrix
R	spin permutation operator
s_A	valence s orbital of atom A
$S_A^2(0)$	s electron density at nucleus A
S	electron spin
$S_{\mu\nu}$	overlap integral between orbitals μ and ν
T	temperature
T_1	spin–lattice relaxation time
T_2	spin–spin relaxation time
$W_{1/2}$	full width (in hertz) of a resonance line at half vertical height
Z_A	atomic number of atom A
α	(i) exchange integral in VB theory; (ii) a spin eigenfunction for a spin 1/2 nucleus
β	(i) Hückel resonance integral; (ii) a spin eigenfunction for a spin 1/2 nucleus
γ	nuclear magnetogyric ratio
δ_a	chemical shift of nucleus a, in parts per million
δ_{ij}	Kronecker delta ($=1$ if $i=j$; $=0$ otherwise)
$\delta(r_{KN})$	Dirac delta operator
ΔE	value of average excitation energy
$\Delta\sigma$	(i) anisotropy of σ; (ii) ring current shielding effect
ε	dielectric constant of medium
η	asymmetry factor in nuclear quadrupole coupling
θ_μ	gauge-invariant atomic orbital μ
Ψ_n	eigenfunction corresponding to E_n
Ψ_n^0	unperturbed Hamiltonian eigenfunction
$\Pi_{s_N s_N}$	mutual polarizability for s electrons on atom N
∇	Laplacian operator
μ	magnetic dipole moment
μ_0	permeability of a vacuum
μ_B	Bohr magneton
μ_N	nuclear magneton

SYMBOLS AND ABBREVIATIONS

ν_i	Larmor precession frequency of nucleus i (in hertz)
ν_0	spectrometer operating frequency
ν_1	frequency of observing r.f. field
σ_i	shielding parameter of nucleus i (in parts per million)
$\sigma_\parallel, \sigma_\perp$	components of σ parallel and perpendicular to a molecular symmetry axis
σ^d	diamagnetic contribution to σ
σ^p	paramagnetic contribution to σ
φ_m	basis function
τ_0	correlation time, general
τ_c	rotational correlation time
τ_{SR}	spin–rotation correlation time
χ_j	atomic orbital j
χ	nuclear quadrupole coupling constant
ω	carrier frequency (in radians per second)
$\omega_i, \omega_0, \omega_1$	as for ν_i, ν_0, ν_1 but in radians per second

ABBREVIATIONS

AEE	average excitation energy approximation
CNDO	complete neglect of differential overlap
CNDO/S	CNDO with spectroscopically derived parameters
CSA	chemical shielding anisotropy
FPT	finite perturbation theory
FT	Fourier transform
GIAO	gauge-invariant atomic orbitals
GTO	Gaussian-type orbital
INDO	intermediate neglect of differential overlap
INDO/S	INDO with spectroscopically derived parameters
LCAO	linear combination of atomic orbitals
MINDO	modified INDO
MO	molecular orbital
NMR	nuclear magnetic resonance
p.p.m.	parts per million
r.f.	radio-frequency
SCF	self-consistent field
SCP	self-consistent perturbation theory
SOS	sum over states
STO	Slater-type orbital
VB	valence bond

1
Introduction to NMR Parameters

In common with other forms of molecular spectroscopy, high resolution NMR gives spectra consisting of a number of lines and bands whose frequency, relative intensity and shape may be analysed to yield molecular parameters. The parameters in question being σ, which describes nuclear shielding; J, which relates to nuclear spin–spin coupling; and the times T_1 and T_2, which concern the relaxation processes encountered by the nuclei which are excited in the NMR experiment. The interpretation of these parameters provides considerable structural and reactivity information which is of interest in many areas of chemistry and chemical physics.

It is our aim to provide the groundwork for an understanding of these NMR parameters in terms of chemically useful information. Hence there is no intention to present a survey of the extensive and rapidly growing NMR literature. Rather, the purpose is to provide a guide to some of the fundamental research publications in this area. There are a number of instructive background texts[1-6] on NMR and several review series [7-12] which serve to present the current status on developments in various areas of NMR spectroscopy.

The nuclear shielding and spin–spin coupling interactions are usually interpreted within a quantum chemical framework. These points are developed later. However, nuclear relaxation interactions are normally considered on the basis of quasi-classical mechanics. The relevent discussion is given in the present chapter. Before turning to the NMR parameters some general remarks on NMR and the evaluation of parameters from high resolution spectra appear to be in order.

1.A GENERAL REMARKS ON NMR

The NMR experiment depends upon the presence of a nucleus having spin. Nuclear spin is described by the quantum number I, which may have values that are positive, integral or half-integral, or zero. The last case arises when both the charge and mass numbers of a nucleus are even, e.g. ^{12}C, ^{16}O and ^{32}S. Thus these particular isotopes are not amenable to NMR study.

Since all nuclei are positively charged there is a magnetic moment, μ_I, collinear with the vector describing the nuclear spin angular momentum. The value of μ_I is given by

$$\mu_I = \gamma\hbar[I(I+1)]^{1/2}, \qquad (1.1)$$

where γ is the magnetogyric ratio, which is a characteristic constant of the nucleus in question. A positive value of γ denotes that the nuclear magnetic moment and spin angular momentum vectors are parallel, whereas a negative value indicates that they are antiparallel.

1.A.1 Resonance Condition

When placed in a magnetic field, B_0 (normally between 1 and 10 T in strength), a nucleus with non-zero spin has the orientation of its spin axis quantized into $2I + 1$ levels whose energies, E, relative to that in zero magnetic field, are given by

$$E = -\gamma\hbar B_0 m_I, \qquad (1.2)$$

where m_I is the magnetic quantum number which may have the values

$$I, \quad I-1, \quad I-2, \quad \ldots, \quad -I+2, \quad -I+1, \quad -I. \qquad (1.3)$$

Hence there is a constant separation, ΔE, between adjacent energy levels, where

$$\Delta E = \gamma\hbar B_0. \qquad (1.4)$$

The selection rule governing magnetic dipole transitions is

$$\Delta m_I = \pm 1; \qquad (1.5)$$

thus NMR occurs when an oscillating external magnetic field, B_1, is applied with the correct polarization and which satisfies the frequency condition

$$\hbar\omega_0 = \Delta E = \gamma\hbar B_0. \qquad (1.6)$$

Hence

$$\omega_0 = \gamma B_0. \qquad (1.7)$$

Equation (1.7) gives the resonance condition where ω_0 is the angular frequency of the radiation from B_1 which is absorbed by the resonating nuclei. It is clear that the frequency at which a given nucleus resonates at constant applied field depends linearly upon its value of γ, which normally appears in the range from 0.4582×10^7 to 28.5335×10^7 rad T^{-1} s^{-1}.

1.A.2 Experimental Aspects

The widespread application of Fourier transform (FT) NMR instruments together with advances in experimental techniques has led to studies on nuclei from most areas of the periodic table. Of the 82 elements up to and including lead only four, Tc, Pm, Ar and Ce, do not have suitable stable isotopes for NMR.[5]

In the past a restrictive influence on NMR studies has been solubility, since high resolution NMR spectra are normally obtained from solutions in inert solvents. However, recent instrumental and experimental developments have rendered it possible to obtain high resolution NMR spectra from a wide variety of solid materials.[13] There is every indication that access to these techniques will become more generally available in the future.

The remaining major limits on the applicability of NMR appear to be low signal intensities and the availability of diamagnetic samples. The signal intensity depends upon the applied field strength as well as on the nuclear properties of isotopic abundance and magnetogyric ratio. Thus, isotopic enrichment and/or higher applied magnetic fields should be conducive to expanding the frontiers of NMR applications.[14]

Paramagnetism tends to induce very large changes in the resonance condition and very broad signals such that the resonance may become unobservable. In the sense that these are thought to be undesirable features of an NMR sample, they are most satisfactorily dealt with by means of chemical modification of the sample. Moreover, there are many instances when the presence of a paramagnetic centre can help to reveal chemically interesting NMR information which might not otherwise be available.[15]

Thus NMR is potentially suitable for investigating the vast majority of molecules and materials. The most popular nuclei, from the NMR standpoint, tend to belong to the first two rows of the periodic table, e.g. those of hydrogen, boron, carbon, nitrogen, fluorine, and phosphorus. However, there appears to be no difference in principle between the molecular factors which determine the NMR parameters of these nuclei and those of heavier nuclei. Consequently most of the examples quoted in this book refer to the more popular nuclei. Another important factor is that quantum chemical calculations at both the *ab initio* and semi-empirical levels are more readily performed, and are less expensive in computer time, for molecules containing light rather than heavy atoms.

1.A.3 Spectral Analysis

When shielding differences between nuclei are much larger than their corresponding spin–spin coupling interactions, the resulting NMR spectra are first order. Under these circumstances nuclear shielding and spin–spin coupling parameters are directly obtained from the positions of the resonance signals. This situation obtains for the majority of chemical environments of most nuclei other than those of hydrogen. The rather unusual position of the hydrogen nuclei arises from their relatively small shielding sensitivity to environmental changes. Hence complicated second-order effects often appear in the spectra of hydrogen nuclei.

If the magnetic field $\mathbf{B}_0$ is applied in the $+z$ direction, then it is possible to construct a spin Hamiltonian operator, $\mathscr{H}_s$, such that

$$\mathscr{H}_s = -\sum_A \gamma_A B_0 \hbar (1 - \sigma_A) I_{AZ} + \sum_{A<B} h J_{AB} \mathbf{I}_A \cdot \mathbf{I}_B, \qquad (1.8)$$

where σ_A is the shielding parameter for nucleus A and J_{AB} is the spin–spin coupling constant for nuclei A and B.

In order to obtain nuclear shielding and spin–spin coupling data from a second-order spectrum it is usual to express $\mathscr{H}_s$ in terms of an eigenvalue problem and then solve it.[1,3] Energy differences between appropriate eigenvalues yield transition frequencies, whereas transition probabilities are obtained from a consideration of the coefficients comprising the eigenfunctions. The predicted spectra are dependent, both in terms of line position and relative intensity, upon the values of σ and J. Thus a comparison between the simulated and experimental spectra, the former of which is varied to obtain an acceptable fit, reveals the appropriate σ and J data.

A similar technique may be used for the analysis of first-order multispin spectra which are complicated by the presence of a large number of chemically distinct nuclear environments. ^{19}F NMR spectra often provide examples of this type of application.

Iterative computer programs for the analysis of complicated multispin spectra, of first or second order as well as in the presence of chemical exchange, are readily available. In Britain they may be obtained from the SERC NMR Computer Program Library at the Daresbury Laboratory, Warrington, and in the USA from the Quantum Chemistry Program Exchange (QCPE) at Bloomington, Indiana.

1.B NUCLEAR SHIELDING

The resonance condition, expressed by equation (1.7), is not of great chemical interest in that it reveals that all nuclei of a given value of γ are expected to produce a single absorption in the NMR spectrum. That this need not be the case was demonstrated in 1950 when two well separated ^{14}N NMR signals were observed for the two nitrogen environments in ammonium nitrate,[16] thus indicating that the signal position depends upon the chemical environment of the nucleus.

As shown by the first term on the right-hand side of equation (1.8), the resonance frequency is related to the shielding of the nucleus, such that equation (1.7) becomes

$$\omega_A = \gamma_A B_0 (1 - \sigma_A). \qquad (1.9)$$

In order to obtain a reliable value for σ_A it is necessary that both ω_A and B_0 be measured to a high accuracy. In practice it is not normally feasible to measure B_0 to the required accuracy, whereas it is possible to measure small differences in resonance frequency at constant applied field. Thus nuclear resonance frequencies are normally measured relative to that of a given nucleus in a standard molecule added to the experimental sample as a reference. The shielding difference, or chemical shift δ, is reported in parts per million, where

$$\delta = (\sigma_{\text{ref}} - \sigma_{\text{sample}}) \times 10^6. \qquad (1.10)$$

Hence a shift of resonance to high frequency, denoted by an increase in δ, corresponds to a decrease in σ_{sample}.

In seeking a molecular interpretation of σ it is necessary to realize that, in a uniform magnetic field $\mathbf{B}_0$, the electrons surrounding a nucleus produce a secondary field, $\mathbf{B}'$, at that nucleus. $\mathbf{B}_0$ and $\mathbf{B}'$ need not be parallel to each other but, as shown by equation (1.11), are related by the second rank tensor $\boldsymbol{\sigma}$:

$$\mathbf{B}' = -\boldsymbol{\sigma} \mathbf{B}_0. \qquad (1.11)$$

In high resolution NMR experiments rapid, and random, motion ensures that the nuclear shielding experienced, σ, is given by one-third of the trace of $\boldsymbol{\sigma}$.

In principle, measurements on liquid crystal and solid phases can yield values for the individual components of the shielding tensor and its anisotropy, $\Delta\sigma$.

For linear and symmetric top molecules,

$$\Delta\sigma = \sigma_\parallel - \sigma_\perp, \qquad (1.12)$$

where $\sigma_\parallel$ refers to the shielding component along the major molecular axis and $\sigma_\perp$ is that in the direction perpendicular to it. The corresponding value of σ is given by

$$\sigma = \tfrac{1}{3}(\sigma_\parallel + 2\sigma_\perp). \qquad (1.13)$$

For less symmetrical molecules,

$$\Delta\sigma = \sigma_{\alpha\alpha} - \tfrac{1}{2}(\sigma_{\beta\beta} + \sigma_{\gamma\gamma}), \qquad (1.14)$$

where the σ_{ii} are the principal tensor components taken in accordance with the convention $\sigma_{\alpha\alpha} \geqslant \sigma_{\beta\beta} \geqslant \sigma_{\gamma\gamma}$. The value of σ is now obtained from

$$\sigma = \tfrac{1}{3}(\sigma_{\alpha\alpha} + \sigma_{\beta\beta} + \sigma_{\gamma\gamma}). \tag{1.15}$$

As shown by equation (1.10), chemical shifts are given as shielding differences. Thus in comparing experimental and theoretical values of chemical shift trends for a given type of nucleus some of the inaccuracies, arising from the incorporation of approximations in the theoretical model, may cancel.[17] A more thorough examination of a theoretical treatment is obtained from a comparison of calculated and experimental values of σ and $\Delta\sigma$.

1.C NUCLEAR SPIN–SPIN COUPLING

Spin–spin couplings, like nuclear shieldings, depend upon the chemical environment of the nuclei concerned. Consequently they are of use in molecular structure determinations. Unlike nuclear shieldings, spin–spin couplings are obtained directly from an NMR spectrum and are independent of the value of the applied field. They are given in hertz, as implied by the formulation of the second term on the right-hand side of equation (1.8).

Nuclear spin–spin coupling results from indirect interactions between the spins of neighbouring nuclei. The nuclear interactions can be transmitted via both the bonding and non-bonding valence electrons. For nuclei A and B the energy of the coupling interaction between them, E_{AB}, is given by

$$E_{AB} = hJ_{AB}I_A \cdot I_B. \tag{1.16}$$

Like σ, J_{AB} is a scalar quantity; an estimate of the anisotropy of the corresponding second rank tensor **J** may be forthcoming from measurements on oriented samples.

Both σ and J are determined by the electronic environments of the nuclei involved. Studies of these parameters for a number of nuclei in a given molecule can provide intimate knowledge of the electronic structure of that molecule. Thus a satisfactory theoretical description of the distribution of electrons in a molecule can lead to reliable predictions of σ and J. Such predictions have a number of chemical applications, such as the identification of the conformation or structure of the species present in a given sample.

Molecular electronic structure is normally described within the framework of quantum chemistry. The most widely applicable of current theories being that of molecular orbitals (MO). Valence bond (VB) theory also has some application to molecular problems. Both of these theories are developed in

Chapter 2 and applied to electronic descriptions of nuclear shielding and spin–spin couplings in Chapters 3 and 4 respectively.

MO computer programs, of various levels of sophistication, are available from the Quantum Chemistry Program Exchange (QCPE). Some of these programs are specifically adapted for calculating spin–spin couplings, the latter are also available from the NMR Computer Program Library at Daresbury.

However, programs for calculating nuclear shielding do not appear to be so readily available. Thus Chapter 5 is devoted to listings of programs, developed in our laboratories, for calculating nuclear shielding using the two commonly encountered perturbation techniques of sum over states (SOS) and finite perturbation theory (FPT). Included in these listings are routines for estimating the effect of the medium on nuclear shieldings. These are based on the solvaton theory discussed in Chapter 3.

1.D NUCLEAR SPIN RELAXATION

The third type of chemically interesting NMR parameter is the time required for nuclear spin relaxation to occur. Relaxation involves the dissipation of excess energy by the excited nuclear spins in order that they may return to the ground state. Since NMR occurs in the radio-frequency region of the electromagnetic spectrum it involves rather low energy transitions; consequently spontaneous emission tends to be of negligible importance for NMR relaxation.

Even in simple cases nuclear spin relaxation may be characterized by two relaxation times, T_1 and T_2. The spin–lattice relaxation time, T_1, relates to the exchange of nuclear magnetization in a direction parallel to that of $\mathbf{B}_0$ and T_2, the spin–spin relaxation time, applies to magnetization in directions perpendicular to $\mathbf{B}_0$.

If inhomogeneity in $\mathbf{B}_0$ is negligible, the NMR lineshape is Lorentzian and its full width at half-height, $w_{1/2}$, is controlled by T_2:

$$w_{1/2} = 1/\pi T_2. \tag{1.17}$$

For non-viscous fluids T_1 and T_2 are usually equal. Thus in the discussion which follows most of the remarks made about T_1 apply equally to T_2.

As well as providing some information on molecular structure, a study of nuclear relaxation times can yield molecular dynamic data[18] and be of assistance in selecting the optimum conditions for the observation of NMR signals from difficult nuclei such as ^{15}N.[19]

There are a number of mechanisms, m, which may contribute to nuclear spin relaxation times. In simple cases the spin relaxation rates due to these mechanisms are additive:

$$\frac{1}{T_1} = \sum_m \frac{1}{T_{1m}}. \qquad (1.18)$$

For any mechanism to be operative in producing spin relaxation it must be responsible for the presence of an oscillating magnetic field at the nuclear site. The frequency of this local field, B_L, needs to be equal to the resonance frequency of the nucleus to be relaxed. If these conditions are fulfilled, then a relaxation transition may be induced.

Before dealing with the processes which produce the requisite local magnetic fields it is appropriate to consider the microdynamic behaviour of molecules in fluids since the fluctuations of these fields are attributed to Brownian motion.

The frequency distribution of the components of B_L is expressed by the power spectral density, $J(\omega)$, which is the Fourier transform of the function describing the various molecular motions in a fluid. Thus $J(\omega_0)$, the component at the resonance frequency, is responsible for nuclear spin–lattice relaxation. The relationship between $J(\omega_0)$ and the ensemble average of B_L is given by

$$J(\omega_0) = 2\langle B_L^2 \rangle_0 \left(\frac{\tau_0^2}{1 + \omega_0^2 \tau_0^2} \right), \qquad (1.19)$$

where τ_0 is the correlation time characterizing the appropriate molecular motions.

The magnitude of $J(\omega_0)$ together with the energy of the interactions between the spin system and the molecular motions determines the value of T_1, as shown by

$$\frac{1}{T_1} = \frac{\mu_0 \gamma^2 \langle B_L^2 \rangle_0 \tau_0}{6\pi(1 + \omega_0^2 \tau_0^2)}. \qquad (1.20)$$

Regarded as a function of τ_0, T_1 has a minimum value when $\omega_0 \tau_0 = 1$, as shown in Fig. 1.1.

In discussing nuclear relaxation phenomena it is normally assumed that the motional narrowing limit applies:

$$\omega_0^2 \tau_0^2 \ll 1. \qquad (1.21)$$

For this to be a reasonable assumption the molecules in question must be tumbling rapidly. This implies small molecules in a low viscosity medium and a relatively high temperature. Under these conditions T_1 becomes frequency independent and equal to T_2, as shown in Fig. 1.1.

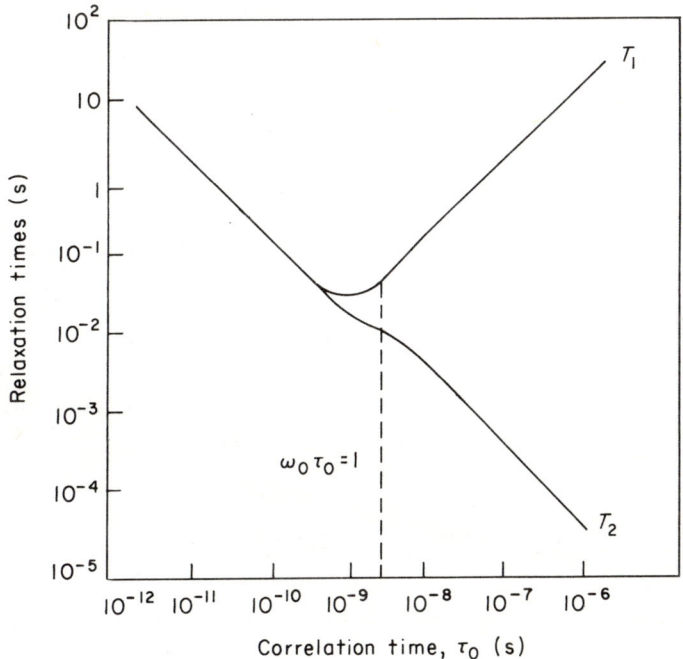

Fig. 1.1 Schematic representation of the nuclear relaxation times, T_1 and T_2, as a function of the correlation time τ_0.

However, for larger molecules, equation (1.21) may not be applicable. This situation is likely to arise in the study of macromolecules, which is becoming of increasing importance. For such molecules T_1 and T_2 are almost certain to be unequal and to have different frequency dependencies since in addition to those occurring at ω_0, low frequency motions also contribute[2] to relaxation processes described by T_2.

Contributions to T_1 can arise from the translational and rotational motions of molecules whereas vibrations are usually too rapid to make effective relaxation pathways. Having considered the frequency characteristics of B_L it is now appropriate to enquire further into the various mechanisms which may contribute to B_L.

The mechanisms are dependent upon various molecular parameters which may thus be studied by means of experimental T_1 values. Additionally, the estimation of the contributing mechanisms is a necessary precondition for calculating an effective correlation time. In the ensuing discussion it is assumed that the extreme narrowing condition obtains. Thus the comments made are those appropriate to the left-hand side of Fig. 1.1.

1.D.1 Relaxation by Magnetic Dipole Interactions

Nuclear magnetic dipole relaxation interactions may occur with other nuclei or with unpaired electrons. These processes usually dominate the relaxation of spin one-half nuclei.

(a) Intramolecular Interactions

If a nucleus I is relaxed by a nucleus of spin S at a distance r, the intramolecular dipole–dipole relaxation time for I, $T_{1dd}(\text{intra})$, is given by

$$\frac{1}{T_{1dd}(\text{intra})} = \frac{\mu_0^2 \gamma_I^2 \gamma_S^2 \hbar^2 S(S+1)\tau_c}{12\pi^2 r^6}. \qquad (1.22)$$

In the case of dipolar interactions with several spins, S, of the same species at different distances, r_i, from I, equation (1.22) becomes

$$\frac{1}{T_{1dd}(\text{intra})} = \frac{\mu_0^2 \gamma_I^2 \gamma_S^2 \hbar^2 S(S+1)\tau_c}{12\pi^2 \sum_i r_i^6}. \qquad (1.23)$$

If I and S belong to the same nuclear species then equation (1.22) is replaced by

$$\frac{1}{T_{1dd}(\text{intra})} = \frac{\mu_0^2 \gamma_I^4 \hbar^2 I(I+1)\tau_c}{8\pi^2 r^6}. \qquad (1.24)$$

The inverse sixth power dependence upon internuclear separation shown in equations (1.22)–(1.24) ensures that the intramolecular dipole–dipole relaxation process becomes rather inefficient in the absence of directly bonded magnetic nuclei. The large magnetogyric ratio of the proton, exceeded only by that of the triton, coupled with its common molecular occurrence, ensures that dipole–dipole interactions with protons frequently dominate the relaxation of other spin-half nuclei, e.g. ^{13}C, ^{15}N.

The correlation time, τ_c, present in equations (1.22)–(1.24) is that describing molecular rotation. This is most readily discussed by assuming that the rotations occur by a small step Brownian process with many steps per radian. Usually the rotating molecule is represented as a rigid top of known symmetry. By assuming spherical solute molecules of volume V in a medium of viscosity η, τ_c may be expressed by

$$\tau_c = \eta V f / kT, \qquad (1.25)$$

where f is a microviscosity factor, which is usually taken to be around 0.16 for pure liquids. Equation (1.25) shows that an increase in viscosity, and molecular size, leads to an enhanced τ_c, whereas an increase in temperature and decrease in the amount of molecular association lead to a decrease in τ_c.

It must be emphasized that this simple description of dipole–dipole relaxation in terms of a single value of τ_c for all nuclei in a molecule is only valid for molecules which rotate isotropically. An example of such a molecule is adamantane, which has two carbon environments, CH and CH_2, the ^{13}C T_1 values for which are 20.5 s and 11.4 s respectively. Thus the ratio of $T_1(CH):T_1(CH_2)$ is 1.798:1.

By means of equation (1.23) the predicted ratio is 2:1 when only directly bonded protons are considered; when the next nearest neighbour protons are included the ratio becomes 1.82:1. Consequently it appears that dipole–dipole interactions provide the dominant, if not the sole, source of ^{13}C relaxation in adamantane.

Observed deviations from the ratio of predicted T_1 values, controlled by proton dipole–dipole interactions, for two or more nuclei in a given molecule, can be used to imply anisotropic molecular tumbling, internal rotation of particular parts of the molecule, hydrogen bonding or other forms of molecular association, or possibly that dipole–dipole interactions are not mainly responsible for the observed T_1 values. Greater sensitivity is achieved if the T_1 values of nuclei with a large magnetogyric ratio difference are employed, rather than a pair of nuclei of the same species.

A pair of spin-half nuclei which fit these requirements, and which occur in most biochemically interesting molecules, is ^{13}C and ^{15}N, whose magnetogyric ratios are 6.7263×10^7 and -2.7107×10^7 rad T^{-1} s^{-1} respectively. An example of this kind of work is afforded by studies on the cation of m-aminobenzoic acid.[21] Internal rotation of the NH_3^+ group may occur in this cation. If free rotation is present then the dipole–dipole controlled value of T_1 for the ^{15}N nucleus is calculated to be about 10 times as large as that for the ^{13}C nucleus in the *para* position to the amino function. For no NH_3^+ internal rotation the corresponding ratio is 1.7:1. The measured ratio in DMSO solution is about 2.5:1 whereas in CF_3SO_3H, as solvent, the ratio is about 10:1. Rapid internal rotation is thus indicated in the latter case and restricted NH_3^+ rotation in the former.

(b) *Intermolecular Interactions*

Nuclei in peripheral molecular sites may relax preferentially by dipole–dipole interactions with nuclear spins on neighbouring molecules. An example is provided by the ^{15}N relaxation of pyridine. At natural abundance levels the nitrogen atom of pyridine has a high probability of being bonded to non-magnetic ^{12}C atoms. The nitrogen lone-pair electrons are available for intermolecular interactions which can account for 85% of the dipolar contribution to the ^{15}N relaxation.

The intermolecular dipole–dipole relaxation rate is given by

$$\frac{1}{T_{1dd}(\text{inter})} = \frac{\mu_0^2 N_S \gamma_I^2 \gamma_S^2 \hbar^2 S(S+1)}{90 D a}, \quad (1.26)$$

where N_S is the concentration of spins per unit volume, a is the distance of closest nuclear approach and D is the mutual self-diffusion constant of the molecules containing the nuclei concerned.

For a spherical molecule of radius r in a spherical sample of viscosity η, the value of D may be found from

$$D = \frac{kT}{6\pi \eta r f}. \quad (1.27)$$

Comparison of equations (1.25) and (1.27) reveals that the dependence of $T_{1dd}(\text{inter})$ upon molecular size, viscosity and temperature is similar to that of $T_{1dd}(\text{intra})$. However, the dependence of $T_{1dd}(\text{inter})$ upon internuclear separation is much less sensitive than it is for $T_{1dd}(\text{intra})$.

(c) Paramagnetic Interactions

If the spin S in equation (1.26) relates to unpaired electrons then, even when no chemical interactions occur between the molecules containing I and S, the large magnetogyric ratio of the electron (658.268 times that of the proton) may be sufficient to produce significant dipole–dipole relaxation of the nuclear spins, I. This situation is often referred to as outer-sphere relaxation and is applicable to some paramagnetic relaxation reagents.[15]

When the unpaired electron and nuclear spins reside in the same molecular species then the nuclear relaxation time, T_1^e, is given by

$$\frac{1}{T_1^e} = \frac{\mu_0^2 \gamma_I^2 \gamma_S^2 \hbar^2 S(S+1)}{120\pi^2 r^6} \left[\frac{3\tau_p}{1+\omega_I^2 \tau_p^2} + \frac{7\tau_p}{1+\omega_S^2 \tau_p^2} \right]$$

$$+ \frac{\mu_0^2 \gamma_S^2 a_N^2 S(S+1)}{24\pi^2} \left[\frac{\tau_e}{1+\omega_S^2 \tau_e^2} \right]. \quad (1.28)$$

The first term on the right-hand side of equation (1.28) represents the dipole–dipole contribution and the second term gives that arising from the contact interaction which depends upon a_N, the nuclear electron hyperfine coupling constant.

In the extreme narrowing limit the first term of equation (1.28) reduces to equation (1.22). The correlation time τ_p is related to the molecular rotation correlation time, τ_c, the electron spin relaxation time, τ_s, and (for a substrate bound to a paramagnetic shift reagent). τ_m, the mean lifetime of the nuclei in the bound state, by

$$\frac{1}{\tau_p} = \frac{1}{\tau_c} + \frac{1}{\tau_s} + \frac{1}{\tau_m}. \quad (1.29)$$

The correlation time τ_e is given by

$$\frac{1}{\tau_e} = \frac{1}{\tau_s} + \frac{1}{\tau_m}. \quad (1.30)$$

Since $\omega_S^2 \tau_e^2$ is a rather large quantity, the second term of equation (1.28) only makes a significant relaxation contribution when a_N is large. However, the corresponding expression for T_2^e is

$$\frac{1}{T_2^e} = \frac{\mu_0^2 \gamma_S^2 a_N^2 S(S+1)}{48\pi^2} \left[\tau_e + \frac{\tau_e}{1 + \omega_S^2 \tau_e^2} \right], \quad (1.31)$$

which reveals that the contract contribution to $1/T_2^e$ is significantly larger than it is for $1/T_1^e$.

Consequently both dipole–dipole and contact interactions reduce the nuclear relaxation times of paramagnetic entities. The latter contribution being particularly large for T_2. On account of these interactions paramagnetic centres are often employed to reduce embarrasingly long relaxation times and to remove nuclear Overhauser enhancements in cases where they are not required. Thus the operating efficiency of the NMR spectrometer may be increased.[22]

1.D.2 Relaxation by Electric Quadrupole Interactions

All nuclei with $I > \frac{1}{2}$ have electric quadrupole moments. The nuclear quadrupole moment, eQ, may interact with an electric field gradient, eq, to provide a very efficient nuclear relaxation process[22] and thus broad NMR signals.

For molecules undergoing isotropic tumbling, characterized by τ_c, the quadrupolar relaxation time, T_{1_q}, is given by

$$\frac{1}{T_{1_q}} = \frac{3\pi^2 (2I+3)}{10 I^2 (2I-1)} \chi^2 (1 + \tfrac{1}{3}\eta^2) \tau_c, \quad (1.32)$$

where χ is the nuclear quadrupole coupling constant, expressed in hertz by

$$\chi = e^2 Q q_{zz}/h, \quad (1.33)$$

where q_{zz} relates to the largest component of the electric field gradient at the nucleus. The asymmetry parameter, η, measures the deviation of the electronic environment of the nucleus from axial symmetry; η lies in the range $0 \leq \eta \leq 1$ and is given by

$$\eta = (q_{xx} - q_{yy})/q_{zz}, \qquad (1.34)$$

where the principal axes, x, y and z, are chosen such that

$$|q_{xx}| \leq |q_{yy}| \leq |q_{zz}|. \qquad (1.35)$$

Thus for axial symmetry $\eta = 0$ and for spherical symmetry the resultant field gradient at the nucleus vanishes, in which case it is clear from equation (1.32) that quadrupolar relaxation is non-operative.

Townes and Dailey[23] have reported a method for providing a rough estimate of the electric field gradient at a nucleus. This involves the assumption that contributions from s electrons may be ignored due to their spherical distribution; the same applies to closed inner shells. Contributions from neighbouring atoms are neglected by assuming that their nuclei and electrons would have cancelling influences at the nucleus concerned. Thus it is deemed that it is sufficient to consider only the valence p, d and f electrons on the atom in question.

If the field gradient due to a single p electron is eq_p, then the field gradients produced along the z axis by n_x, n_y and n_z electrons in the respective p orbitals are

$$eq_{zz}(p_z) = n_z eq_p, \qquad (1.36)$$

$$eq_{zz}(p_y) = -\tfrac{1}{2} n_y eq_p, \qquad (1.37)$$

$$eq_{zz}(p_x) = -\tfrac{1}{2} n_x eq_p. \qquad (1.38)$$

Similar expressions are available for eq_{xx} and eq_{yy}. From equations (1.36)–(1.38) it follows that the total electric field gradient along the z axis is

$$eq_{zz} = [n_z - \tfrac{1}{2}(n_x + n_y)] eq_p \qquad (1.39)$$

and

$$\eta = \frac{\tfrac{3}{2}(n_x - n_z)}{[n_z - \tfrac{1}{2}(n_x + n_y)]}. \qquad (1.40)$$

A consideration of equations (1.32), (1.33), (1.39) and (1.40) reveals that T_{1q} depends critically upon the electronic environment of the nucleus. This is demonstrated by the ^{35}Cl line widths of ~ 10 Hz for NaCl and ~ 10 kHz for CCl$_4$. In the former case the approximately spherical electronic environment of the chloride ion leads to the absence of appreciable electric quadrupole relaxation, whereby the line width is controlled by the less efficient dipole–dipole process. For covalently bonded CCl$_4$ the large field gradients at the chlorine nuclei cause rapid quadrupolar relaxation.

Equations (1.32), (1.39) and (1.40), in conjunction with MO data, may be useful for assigning the signals of quadrupolar nuclei on the basis of their relative line widths. This approach has been applied successfully to the ^{14}N NMR spectra of some heteroaromatic systems with non-equivalent nitrogen atoms.[24]

The presence of τ_c in equation (1.32) ensures that the dependence of T_{1_q} upon temperature, molecular size and viscosity is the same as that for $T_{1_{dd}}$.

1.D.3 Relaxation by Spin–Rotation Interactions

Spin–rotation relaxation arises from the interaction between nuclear magnetic moments and rotational magnetic moments of the molecules containing those nuclei. Thus a direct transfer occurs of nuclear spin energy to the process of molecular rotation. This contrasts with the dipole–dipole and quadrupole mechanisms which operate via an indirect energy transfer.

By assuming that the motion is isotropic the nuclear spin–rotation relaxation rate, $1/T_{1SR}$, is given by

$$\frac{1}{T_{1SR}} = \frac{2MkT(C_\parallel^2 + 2C_\perp^2)\tau_{SR}}{3\hbar^2}, \quad (1.41)$$

where M represents the molecular moment of inertia, $C_\parallel$ and $C_\perp$ respectively are the components of the spin–rotation tensor parallel to and perpendicular to the molecular symmetry axis and the correlation time, τ_{SR}, is given for a spherical molecule in a diffusion-controlled process by Hubbard's relation:[25]

$$\tau_c \tau_{SR} = M/6kT. \quad (1.42)$$

From a comparison of equations (1.25) and (1.42) it follows that the temperature dependence of τ_{SR} is inversely proportional to that of the viscosity of the medium. Usually η for liquids decreases rapidly as the temperature is increased, thus leading to an increase in τ_{SR} at higher temperatures.

Equation (1.41) reveals that T_{1SR} decreases as the temperature increases, which is in contrast to the other relaxation mechanisms considered here.

The diagonal elements of the spin–rotation tensor become larger as the rate of molecular motion increases. This leads to a more efficient exchange of nuclear spin energy. Thus spin–rotation relaxation is expected to be at its most dominant for small molecules tumbling rapidly at high temperatures. Consequently it is likely to be of particular importance for vapour phase studies.

An example of the operation of the spin–rotation mechanism is afforded by the ^{31}P relaxation of triethylphosphine at 302 K. As the temperature is decreased the spin–rotation process becomes of less importance.[26] At 353 K about 50% of the ^{15}N relaxation rate for *trans*-azobenzene is due to spin–rotation, again the dipole–dipole process becomes of dominating significance at lower temperatures.[27]

For ^{13}C nuclei spin–internal rotation of freely rotating groups, such as methyl groups, can be an important relaxation pathway even when the large molecule to which such groups are attached is not rotating rapidly.

1.D.4 Relaxation by Nuclear Shielding Anisotropy

Brownian motion can modulate the nuclear shielding tensor and thus provide a fluctuating magnetic field. If this field has components in the region of the appropriate resonance frequency then nuclear relaxation may result. The relaxation times, T_{1SA} and T_{2SA}, due to this interaction depend upon the applied magnetic field and the rotational correlation time as follows:

$$\frac{1}{T_{1SA}} = \frac{\mu_0 \gamma_I^2 B_0^2 \Delta\sigma^2 \tau_c}{30\pi} \tag{1.43}$$

and

$$\frac{1}{T_{2SA}} = \frac{7\mu_0 \gamma_I^2 B_0^2 \Delta\sigma^2 \tau_c}{180\pi}. \tag{1.44}$$

As shown by equations (1.43) and (1.44), for the motional rowing limit, $T_{1SA}:T_{2SA}$ is 1:0.857 thus, under these conditions, nuclear shielding anisotropy controls the relaxation by T_2, rather than T_1, processes.

The dependence of T_{1SA} and T_{2SA} upon B_0 means that this relaxation mechanism may be distinguished from others by measurements made as a function of applied field. The dependence also implies that this relaxation process will assume a greater practical importance as the distribution of superconducting NMR magnets becomes more widespread. This seems likely to be particularly true for heavy nuclei, which tend to have larger shielding anisotropies than do lighter nuclei. For example, values of $\Delta\sigma$ ranging from 2500 to 10 500 p.p.m. have been reported for complexes of ^{195}Pt and similar values are available for some ^{199}Hg compounds.[28,29]

In contrast to this the largest nitrogen shielding anisotropy reported to date is 672 ± 20 p.p.m. for pyridine. At 7.42 T this makes a significant contribution to the ^{15}N relaxation rate.[30] Variable temperature and field studies[32] on Pb(CH$_3$)$_3$Cl show that spin–rotation controls the ^{207}Pb relaxation at 2.35 T whereas shielding anisotropy dominates at 7.05 T.

1.D.5 Relaxation by Scalar Coupling

If chemical exchange or internal rotation causes the spin–spin coupling interaction between two nuclei to become time dependent then *scalar*

1 INTRODUCTION TO NMR PARAMETERS

relaxation of the first kind can occur. *Scalar relaxation of the second kind* relates to the case when the relaxation rate of a coupled nucleus is fast compared to $2\pi J$. Coupling to a quadrupolar nucleus can give rise to this relaxation mechanism.

The contribution of scalar coupling to T_1 is given by T_{1sc}, where

$$\frac{1}{T_{1sc}} = \frac{8}{3}\pi^2 J^2 S(S+1)\left[\frac{\tau_1^s}{1+(\omega_I-\omega_S)^2(\tau_1^s)^2}\right], \quad (1.45)$$

the comparable expression for T_{2sc} being

$$\frac{1}{T_{2sc}} = \frac{1}{2T_{1sc}} + \frac{4}{3}\pi^2 J^2 S(S+1)\tau_1^s. \quad (1.46)$$

Equations (1.45) and (1.46) become frequency independent when the motional narrowing limit is reached. In these equations τ_1^s refers to the exchange or rotation lifetime for the first kind of scalar relaxation and to the relaxation time of nucleus S for the second kind.

For scalar coupling relaxation to be operative it is important that the resonance frequencies of the coupled nuclei be similar. Otherwise the denominator of equation (1.45) will be very large, thus rendering the relaxation mechanism rather inefficient. This may be compensated for to some extent by a large value of J.

As shown by equation (1.46) the scalar coupling contribution to T_2 can be important when $1/\tau_1^s$ is comparable to $2\pi J$. This situation is commonly encountered in the NMR spectra of protons spin–spin coupled to quadrupolar nuclei such as ^{14}N. In such cases the T_{2sc} process results in broadened proton signals.

An example of scalar coupling relaxation of the second kind has been reported for a solution of $Tl(CH_3)_2NO_3$ in degassed D_2O at 9.4 T and 300 K.[31] The half-width of the proton signal is 55 Hz, which appears to be due to scalar coupling with ^{205}Tl which relaxes rapidly due to its shielding anisotropy.

An interesting consequence of scalar coupling relaxation depends upon its relative inefficiency. A typical value of τ_0 for a macromolecule might be 10^{-8} s. As shown in Fig. 1.1, this corresponds to a small value of T_1 and thus to a rather broad resonance signal. This dipolar broadening may lead to difficulties in resolving the signals from the various ^{15}N nuclei bonded to protons in a protein. Substitution of deuterons for the protons leads to much narrower ^{15}N signals since the magnetogyric ratio of the deuteron is about one-sixth that of the proton, and thus to greater resolution. However, the deuteron also has a quadrupole moment but its contribution to the ^{15}N line width occurs only via the rather inefficient scalar coupling process which also depends, via J, on the magnetogyric ratio. The outcome being that seven ^{15}N glycyl resonances are determined from a D_2O solution of haemoglobin whereas only three poorly resolved signals appear in the spectrum taken on a H_2O solution.[33]

REFERENCES

1. J. A. Pople, W. G. Schneider and H. J. Bernstein, "High Resolution NMR", McGraw-Hill, New York (1959).
2. A. Abragam, "The Principles of Nuclear Magnetism", Clarendon Press, Oxford (1961).
3. A. Carrington and A. D. McLachlan, "Introduction to Magnetic Resonance", Harper and Row, New York (1967).
4. F. W. Wehrli and T. Wirthlin, "Interpretation of ^{13}C NMR Spectra", Heyden, London (1976).
5. R. K. Harris and B. E. Mann (eds), "NMR and the Periodic Table", Academic Press, London (1978).
6. E. D. Becker, "High Resolution NMR: Theory and Chemical Applications", 2nd edn, Academic Press, New York (1980).
7. "Advances in Magnetic Resonance", Vols 1–9 (J. S. Waugh, ed.), Academic Press, New York.
8. "Annual Reports on NMR Spectroscopy", Vols 1–16 (E. F. Mooney, ed., Vols 1–6; G. A. Webb, ed., Vols 7–16), Academic Press, London.
9. "NMR Basic Principles and Progress", Vols 1–20 (P. Diehl, E. Fluck and R. Kosfeld, eds), Springer-Verlag, Berlin, series discontinued.
10. "Progress in NMR Spectroscopy", Vols 1–14 (J. W. Emsley, J. Feeney and L. H. Sutcliffe, eds), Pergamon, Oxford.
11. "Specialist Periodical Reports on NMR", Vols 1–14 (R. K. Harris, ed., Vols 1–5; R. J. Abraham, ed., Vols 6–8; G. A. Webb, ed., Vols 9–14), Royal Society of Chemistry, London.
12. "Topics in ^{13}C NMR Spectroscopy", Vols 1–3 (G. C. Levy, ed.), John Wiley & Sons, New York.
13. R. E. Wasylishen and C. A. Fyfe, in "Annual Reports on NMR Spectroscopy", Vol. 12 (G. A. Webb, ed.), Academic Press, London (1982), p. 1.
14. T. Axenrod and G. A. Webb (eds), "NMR Spectroscopy of Nuclei Other than Protons", John Wiley & Sons, New York (1974).
15. G. A. Webb, in "Annual Reports on NMR Spectroscopy", Vol. 6A (E. F. Mooney, ed.), Academic Press, London (1975), p. 1.
16. W. G. Proctor and F. C. Yu, *Phys. Rev.*, 77, 717 (1950).
17. K. A. K. Ebraheem and G. A. Webb, in "Progress in NMR Spectroscopy", Vol. 11 (J. Emsley, J. Feeney and L. Sutcliffe, eds), Pergamon Press, Oxford (1977), p. 149.
18. L. M. Jackman and F. A. Cotton (eds), "Dynamic Nuclear Magnetic Resonance Spectroscopy", Academic Press, New York (1975).
19. M. Witanowski, L. Stefaniak and G. A. Webb, in "Annual Reports on NMR Spectroscopy", Vol. 11B (G. A. Webb, ed.), Academic Press, London (1981), p. 1.

20. C. P. Slichter, "Principles of Magnetic Resonance", Springer-Verlag, Berlin (1978), p. 137.
21. G. C. Levy, A. D. Godwin, J. M. Hewitt and C. Sutcliffe, *J. magn. Reson.*, **29**, 553 (1978).
22. M. Witanowski and G. A. Webb (eds), "Nitrogen NMR", Plenum Press, London (1973).
23. C. H. Townes and B. P. Dailey, *J. Chem. Phys.*, **17**, 782 (1949).
24. M. Witanowski, L. Stefaniak and G. A. Webb, *J. magn. Reson.*, **36**, 227 (1979).
25. P. S. Hubbard, *Phys. Rev.*, **131**, 1155 (1963).
26. N. J. Koole, A. J. De Koning and M. J. A. De Bie, *J. magn. Reson.*, **25**, 375 (1977).
27. J. B. Lambert and D. A. Netzel, *J. magn. Reson.*, **25**, 531 (1977).
28. H. J. Keller and H. H. Rupp, *Z. Naturforsch.*, **26a**, 785 (1971).
29. J. D. Kennedy and W. McFarlane, *J. chem. Soc. Faraday II*, **72**, 1653 (1976).
30. D. Schweitzer and H. W. Spiess, *J. magn. Reson.*, **15**, 529 (1974).
31. F. Brady, R. W. Matthews, M. J. Forster and D. G. Gillies, *J. chem. Soc. chem. Commun.*, **1981**, 911 (1981).
32. G. R. Hays, D. G. Gillies, C. P. Blaauw and A. D. H. Claque, *J. magn. Reson.*, **45**, 102 (1981).
33. A. Lapidot and C. S. Irving, *J. Am. chem. Soc.*, **99**, 5488 (1977).

2
Introduction to Quantum Chemistry

If we wish to learn about the NMR properties of a molecule such as nuclear shielding and spin–spin coupling, we may profitably approach the problem as one relating to interactions between nuclei and electrons in the molecule. In which case a knowledge of quantum chemistry can provide a satisfactory account of the subject.

2.A FORMALISM OF QUANTUM CHEMISTRY[1-3]

2.A.1 Schrödinger's Equation

Information on electronic structure and electronic states of atoms and molecules is obtained from solving the time-independent Schrödinger equation,

$$\mathcal{H}\Psi = E\Psi, \tag{2.1}$$

where $\mathcal{H}$ is a quantum mechanical Hamiltonian operator appropriate to atomic and molecular systems; Ψ is a wavefunction describing a stationary state with energy E, where Ψ is a function of the space and spin coordinates of the electrons in the system under consideration. Ψ is often called the eigenfunction and E is the corresponding eigenvalue. The integral of the square of the function $\int \Psi^2 \, dv$, or more precisely $\int \Psi^*\Psi \, dv$, is a measure of the probability of finding electrons in the volume element dv, where Ψ^* is the complex conjugate of Ψ.

The Hamiltonian operator contains expressions relating to the kinetic energy of the nuclei and electrons, and an electrostatic potential energy, such that $\mathcal{H}$ is written as

$$\mathcal{H} = -\frac{\hbar^2}{2m}\sum_i \nabla_i^2 - \frac{\hbar^2}{2}\sum_k \frac{1}{M_k}\nabla_k^2 - \sum_{i>k}\sum \frac{Z_k e^2}{r_{ik}}$$
$$+ \sum_{i>j}\sum \frac{e^2}{r_{ij}} + \sum_{k>l}\sum \frac{Z_k Z_l e^2}{r_{kl}}, \tag{2.2}$$

where $\hbar = h/2\pi$, h being the Planck constant; m and M are the masses of the electron and nucleus, respectively, e the charge of an electron, Z the atomic number, r_{ij} the distance between electrons i and j, r_{ik} the distance between electron i and nucleus k, and r_{kl} the distance between nuclei k and l. The symbol ∇ is the Laplacian, defined as the sum of second derivatives by

$$\nabla^2 = \frac{\partial^2}{\partial x^2} + \frac{\partial^2}{\partial y^2} + \frac{\partial^2}{\partial z^2}. \tag{2.3}$$

Since the mass of a proton is about 1840 times larger than that of an electron, we may use the so-called Born–Oppenheimer approximation which simplifies the Hamiltonian operator given in equation (2.2). The approximation amounts to separating the nuclear kinetic energy and the electrostatic potential energy among the nuclei from the purely electronic term in $\mathcal{H}$. This results in the electronic Hamiltonian operator $\mathcal{H}^{\text{elec}}$ for our further consideration:

$$\mathcal{H}^{\text{elec}} = -\frac{\hbar^2}{2m}\sum_i \nabla_i^2 - \sum_{i>k}\sum \frac{Z_k e^2}{r_{ik}} + \sum_{i>j}\sum \frac{e^2}{r_{ij}}. \tag{2.4}$$

We shall use the symbol $\mathcal{H}$ rather than $\mathcal{H}^{\text{elec}}$ below, since confusion is unlikely to arise from this abbreviation.

Next, we have the problem of obtaining the eigenvalues and eigenfunctions of $\mathcal{H}$. As is well known, we can only obtain the solution of Schrödinger's equation analytically for hydrogen and hydrogen-like atoms. The solution for molecules containing many electrons must be found by appropriate approximations such as those arising in the variation and perturbation methods.

2.A.2 The Variation Method

Our aim is to obtain an approximate eigenfunction, Ψ, in such a way as to make as good an approximation as possible to the exact eigenfunction. The situation presented in equation (2.1) is equivalent to the following mathematical treatment by the variation method. If the variations in Ψ are completely arbitrary, we can determine the best possible Ψ for the chosen Hamiltonian $\mathcal{H}$ using the condition given by equations (2.5) and (2.6) to obtain Ψ such that $\int \Psi^* \mathcal{H} \Psi \, dv$ takes a stationary value together with the auxiliary condition of normalization; $\int \Psi^* \Psi = 1$. We may thus write

$$I(\Psi) \equiv \int \Psi^* \mathcal{H} \Psi \, dv - \lambda \int \Psi^* \Psi \, dv \tag{2.5}$$

and

$$\delta I(\Psi) = 0. \tag{2.6}$$

2 INTRODUCTION TO QUANTUM CHEMISTRY

The variation, $\delta I(\Psi)$, in $I(\Psi)$ allows us to write

$$\delta I(\Psi) = \int \delta\Psi^* \mathcal{H}\Psi \, dv + \int \Psi^* \mathcal{H} \, \delta\Psi \, dv$$
$$- \lambda \int \delta\Psi^* \Psi \, dv - \lambda \int \Psi^* \delta\Psi \, dv$$
$$= \int \delta\Psi^*(\mathcal{H}\Psi - \lambda\Psi) \, dv + \int \delta\Psi(\mathcal{H}\Psi^* - \lambda\Psi^*) \, dv$$
$$= 0, \tag{2.7}$$

where we have used the property that $\mathcal{H}$ is hermitian. Since $\delta\Psi$ and $\delta\Psi^*$ are analytically independent, it follows that

$$\mathcal{H}\Psi - \lambda\Psi = 0 \quad \text{and} \quad \mathcal{H}\Psi^* - \lambda\Psi^* = 0. \tag{2.8}$$

These equations are equivalent to the Schrödinger equation, equation (2.1), if we substitute E in place of λ.

We suppose that the function Ψ can be expanded in terms of a complete set of basis functions χ_i:

$$\Psi = \sum_{m=1}^{s} C_m \chi_m. \tag{2.9}$$

Substituting equation (2.9) into equation (2.5), we have

$$I(\Psi) = \sum_{m,n} C_m^* C_n \langle m|\mathcal{H}|n \rangle - E \sum_{m,n} C_m^* C_n \langle m|n \rangle, \tag{2.10}$$

where

$$\langle m|\mathcal{H}|n \rangle = \int \chi_m^* \mathcal{H} \chi_n \, dv = H_{mn},$$

$$\langle m|n \rangle = \int \chi_m^* \chi_n \, dv = S_{mn}. \tag{2.11}$$

Under the condition that Ψ satisfies equation (2.6), by differentiating equation (2.10) with respect to C_m^*, we have

$$\sum_{n} \{\langle m|\mathcal{H}|n \rangle - E \langle m|n \rangle\} C_n = 0 \tag{2.12}$$

where $m = 1, 2, \ldots, N$. According to a standard theorem in algebra, the required condition that the simultaneous equations (2.12) have a solution other than the trivial one of $C_n = 0$ is that the determinant, called the secular determinant of the coefficients, should vanish: namely

$$\begin{vmatrix} H_{11} - ES_{11} & H_{12} - ES_{12} & \ldots & H_{1N} - ES_{1N} \\ H_{21} - ES_{21} & H_{22} - ES_{22} & \ldots & H_{2N} - ES_{2N} \\ \vdots & \vdots & & \vdots \\ H_{N1} - ES_{N1} & H_{N2} - ES_{N2} & \ldots & H_{NN} - ES_{NN} \end{vmatrix} = 0. \qquad (2.13)$$

As is shown above, these procedures form the basis of the calculation of the most probable eigenfunctions and eigenvalues.

2.A.3 Perturbation Methods

Another way of obtaining an approximate solution is the perturbation method. We may use this method when we want to obtain the eigenvalues and eigenfunctions for a perturbed Hamiltonian operator using eigenvalues and eigenfunctions for an unperturbed Hamiltonian operator.

It becomes necessary to solve the problem for the perturbed Hamiltonian operator $\mathcal{H}$ whose Schrödinger equation is

$$\mathcal{H} = \mathcal{H}_0 + \lambda \mathcal{H}', \qquad (2.14)$$

where $\mathcal{H}_0$ is the unperturbed Hamiltonian operator and $\lambda \mathcal{H}'$ is a perturbation term, which is small in magnitude compared with $\mathcal{H}_0$. λ is used as a real parameter for convenience and is small with respect to unity. If the eigenvalues E_n^0 and the eigenfunctions Ψ_n^0 for the unperturbed Hamiltonian operator are given by

$$\mathcal{H}_0 \Psi_n^0 = E_n^0 \Psi_n^0, \qquad (2.15)$$

then the eigenvalues E_n and eigenfunctions Ψ_n for the perturbed Hamiltonian are given by

$$(\mathcal{H}_0 + \lambda \mathcal{H}') \Psi_n = E_n \Psi_n. \qquad (2.16)$$

When λ approaches zero, E_n and Ψ_n approach E_n^0 and Ψ_n^0, respectively. Therefore, within the limitation of small values of λ, E_n and Ψ may be approximated as

$$E_n = E_n^0 + \lambda E_n' + \lambda^2 E_n'' + \ldots, \qquad (2.17)$$

$$\Psi_n = \Psi_n^0 + \lambda \Psi_n' + \lambda^2 \Psi_n'' + \ldots, \qquad (2.18)$$

where $\lambda E_n'$ and $\lambda \Psi_n'$ are referred to as the first-order perturbation terms and $\lambda^2 E_n''$ and $\lambda^2 \Psi_n''$ as the second-order perturbation terms, etc. By substituting equations (2.17) and (2.18) into equation (2.1), we have

$$\mathcal{H}_0 \Psi_n^0 + \lambda(\mathcal{H}' \Psi_n^0 + \mathcal{H}_0 \Psi_n') + \lambda^2(\mathcal{H}' \Psi_n' + \mathcal{H}_0 \Psi_n'') + \ldots$$
$$= E_n^0 \Psi_n^0 + \lambda(E_n' \Psi_n^0 + E_n^0 \Psi_n') + \lambda^2(E_n^2 \Psi_n^0 + E_n' \Psi_n' + E_n^0 \Psi_n'') + \ldots. \qquad (2.19)$$

We may attempt to solve equation (2.19) by equating the coefficients of successive powers of λ to zero. In doing so we obtain the following set of equations:

For λ^0, $\quad \mathcal{H}_0 \Psi_n^0 = E_n^0 \Psi_n^0.$ \hfill (2.20)

For λ^1, $\quad (\mathcal{H}_0 - E_n^0)\Psi_n' = E_n' \Psi_n^0 - \mathcal{H}' \Psi_n^0.$ \hfill (2.21)

For λ^2, $\quad (\mathcal{H}_0 - E_n^0)\Psi_n'' = E_n'' \Psi_n^0 + E_n' \Psi_n' - \mathcal{H}' \Psi_n'.$ \hfill (2.22)

Equations (2.20)–(2.22) may be solved by means of expanding the functions Ψ_n^i in terms of the functions of the unperturbed Hamiltonian operator:

$$\Psi_n' = \sum_{m \neq k} C_m \Psi_m^0. \tag{2.23}$$

By substituting equation (2.23) into equations (2.20)–(2.22), we have

$$\sum_m C_m (\mathcal{H}^0 - E_k^0)\Psi_m^0 = -(\mathcal{H}' - E_k')\Psi_k^0. \tag{2.24}$$

If we multiply both sides of equation (2.24) from the left by Ψ_n^{0*}, impose the condition $\langle \Psi_k^0 | \Psi_k^n \rangle = 0$, and then integrate, we obtain

$$C_n(E_n^0 - E_k^0) = -\langle \Psi_n^0 | \mathcal{H}' | \Psi_k^0 \rangle. \tag{2.25}$$

Thus, we have the following first-order perturbed eigenfunction by using equations (2.23) and (2.25):

$$\Psi_k' = -\sum_{m \neq k} \frac{\langle \Psi_m^0 | \mathcal{H}' | \Psi_k^0 \rangle}{E_m^0 - E_k^0} \Psi_m^0. \tag{2.26}$$

Substituting equation (2.26) into equations (2.20)–(2.22), we have the following second-order perturbed eigenvalue expression:

$$E_k'' = -E_k' \langle \Psi_k^0 | \Psi_k' \rangle + \langle \Psi_k^0 | \mathcal{H}' | \Psi_k' \rangle. \tag{2.27}$$

Then, substitution of equation (2.26) into equation (2.27) gives

$$E_k'' = -\sum_{m \neq k} \frac{\langle \Psi_k^0 | \mathcal{H}' | \Psi_m^0 \rangle \langle \Psi_m^0 | \mathcal{H}' | \Psi_k^0 \rangle}{E_m^0 - E_k^0}. \tag{2.28}$$

Pople et al.[4] have proposed a general method for the calculation of self-consistent MO wavefunctions in the presence of small, but finite, perturbations. We will attempt to introduce the finite perturbation theory (FPT) method, which is very useful for the calculation of nuclear shieldings and spin–spin couplings.

In the FPT approach it is assumed that the energy can be expanded as a Taylor series in powers of the parameter $\lambda_r, r = 1, 2, \ldots$:

$$E(\lambda) = E^0 + \sum_r \lambda_r E_r' + \tfrac{1}{2} \sum_r \sum_s \lambda_r \lambda_s E_{rs}'' + \ldots, \tag{2.29}$$

where

$$E'_r = \left(\frac{\partial E(\lambda)}{\partial \lambda_r}\right)_0, \quad (2.30)$$

$$E''_{rs} = \left(\frac{\partial^2 E(\lambda)}{\partial \lambda_r \partial \lambda_s}\right)_0, \quad (2.31)$$

$$\lambda = (\lambda_r, \lambda_s, \ldots). \quad (2.32)$$

Here E^0 is the energy in the absence of any perturbation. Various physical properties depend on the perturbation associated with the individual powers of λ_r. In addition, the wavefunction may be expanded in a Taylor series similar to equation (2.29) as

$$\Psi(\lambda) = \Psi^0 + \sum_r \lambda_r \Psi'_r + \ldots, \quad (2.33)$$

where

$$\Psi'_r = \left(\frac{\partial \Psi(\lambda)}{\partial \lambda_r}\right)_{\lambda=0}.$$

Here, Ψ^0 is the wavefunction in the absence of any perturbation. Ψ'_r is the first-order correction to Ψ^0.

If the perturbed Hamiltonian is given as

$$\mathcal{H}(\lambda) = \mathcal{H}_0 + \sum_r \lambda_r \mathcal{H}_r, \quad (2.34)$$

the second-order perturbed eigenvalue E''_{rs} is represented as

$$E''_{rs} = \langle \Psi^0 | \mathcal{H}_r | \Psi'_s \rangle$$
$$= \langle \Psi^0 | \mathcal{H}_r | (\partial \Psi / \partial \lambda_s)_{\lambda=0} \rangle. \quad (2.35)$$

In the finite perturbation treatment the wavefunction $\Psi(\lambda)$ is calculated for certain finite values of the perturbing field λ. Of course, in the limit of small perturbations, the result is equivalent to a complete solution of the infinitesimal perturbation problem. However, there is a resulting gain in computational efficiency. Let us proceed to develop the FPT method.

The energy, $E(\lambda)$, is given as

$$E(\lambda) = \langle \Psi(\lambda) | \mathcal{H}(\lambda) | \Psi(\lambda) \rangle. \quad (2.36)$$

2 INTRODUCTION TO QUANTUM CHEMISTRY

The first derivative of $E(\lambda)$ becomes

$$\frac{\partial E(\lambda)}{\partial \lambda} = \left\langle \frac{\partial \Psi(\lambda)}{\partial \lambda} \middle| \mathcal{H}(\lambda) \middle| \Psi(\lambda) \right\rangle$$
$$+ \left\langle \Psi(\lambda) \middle| \frac{\partial \mathcal{H}(\lambda)}{\partial \lambda} \middle| \Psi(\lambda) \right\rangle$$
$$+ \left\langle \Psi(\lambda) \middle| \mathcal{H}(\lambda) \middle| \frac{\partial \Psi(\lambda)}{\partial \lambda} \right\rangle. \tag{2.37}$$

Since the Hamiltonian $\mathcal{H}(\lambda)$ is hermitian, we obtain the following relation:

$$\left\langle \Psi(\lambda) \middle| \mathcal{H}(\lambda) \middle| \frac{\partial \Psi(\lambda)}{\partial \lambda} \right\rangle = \left\langle \frac{\partial \Psi(\lambda)}{\partial \lambda} \middle| \mathcal{H}(\lambda) \middle| \Psi(\lambda) \right\rangle. \tag{2.38}$$

Using

$$\mathcal{H}(\lambda)\Psi(\lambda) = E(\lambda)\Psi(\lambda), \qquad \mathcal{H}(\lambda)^*\Psi(\lambda)^* = E(\lambda)\Psi(\lambda)^*$$

and

$$\langle \Psi(\lambda)|\Psi(\lambda)\rangle = \langle \Psi(\lambda)|\Psi(\lambda)\rangle = 1,$$

we obtain

$$\frac{\partial E(\lambda)}{\partial \lambda} = \left\langle \Psi(\lambda) \middle| \frac{\partial \mathcal{H}(\lambda)}{\partial \lambda} \middle| \Psi(\lambda) \right\rangle$$
$$+ E(\lambda) \left\{ \left\langle \frac{\partial \Psi(\lambda)}{\partial \lambda} \middle| \Psi(\lambda) \right\rangle + \left\langle \Psi(\lambda) \middle| \frac{\partial \Psi(\lambda)}{\partial \lambda} \right\rangle \right\}$$
$$= \left\langle \Psi(\lambda) \middle| \frac{\partial \mathcal{H}(\lambda)}{\partial \lambda} \middle| \Psi(\lambda) \right\rangle$$
$$+ E(\lambda) \frac{\partial}{\partial \lambda} \{\langle \Psi(\lambda)|\Psi(\lambda)\rangle + \langle \Psi(\lambda)|\Psi(\lambda)\rangle\}$$
$$= \left\langle \Psi(\lambda) \middle| \frac{\partial \mathcal{H}(\lambda)}{\partial \lambda} \middle| \Psi(\lambda) \right\rangle. \tag{2.39}$$

This relation is the so-called Hellmann–Feymann theorem. If the Hamiltonian $\mathcal{H}(\lambda)$ has a linear form such as equation (2.34), then equation (2.39) becomes

$$\frac{\partial E(\lambda)}{\partial \lambda} = \langle \Psi(\lambda)|\mathcal{H}_r|\Psi(\lambda)\rangle. \tag{2.40}$$

If the right-hand side is written as $f_r(\lambda)$, which becomes the expectation value of $\mathcal{H}_r$, then the first derivative of $f_r(\lambda)$ is related to equation (2.31) by

$$\frac{\partial^2 E(\lambda)}{\partial \lambda_r \partial \lambda_s} = \frac{\partial f_r}{\partial \lambda_s}.$$

Taking the derivatives at $\lambda_1 = \lambda_2 = \ldots = 0$, we obtain

$$E''_{rs} = \left(\frac{\partial f_r(\lambda)}{\partial \lambda_s}\right)_{\lambda=0} \qquad (2.41)$$

As stated above, the Hellmann–Feynmann theorem enables us to calculate a first derivative of $f_r(\lambda)$ rather than a second derivative of energy $E(\lambda)$ explicitly.

2.B MOLECULAR ORBITAL THEORY[5,6]

Electrons in a molecule are associated with nuclei and instead of the original atomic orbitals (AOs; 1s, 2s, 2p), on each atom, we have to consider molecular orbitals (MOs), each of which consists of a linear combination of AOs (LCAO). For many electron molecules this provides a sufficiently good approximate description to be generally useful.

2.B.1 Wavefunction Ψ

Formulation of many electron wavefunctions is carried out on the basis of the product of orthonormalized one-electron wavefunctions in a molecule. The wavefunction Ψ of a system containing $2n$ electrons is expressed by a Slater determinant:

$$\Psi = \frac{1}{\sqrt{2n!}} \begin{vmatrix} \varphi_1(1)\alpha(1) & \varphi_1(2)\alpha(2) & \ldots & \varphi_1(2n)\alpha(2n) \\ \varphi_2(1)\beta(1) & \varphi_2(2)\beta(2) & \ldots & \varphi_2(2n)\beta(2n) \\ \vdots & \vdots & & \vdots \\ \varphi_n(1)\beta(1) & \varphi_n(2)\beta(2) & \ldots & \varphi_n(2n)\beta(2n) \end{vmatrix}, \qquad (2.42)$$

where φ_i is the orbital part of the wavefunction, and α or β represent the electronic spin part. Ψ must be antisymmetrical, i.e. it must change its sign upon exchange of the coordinates of any two electrons. This is necessary in order to satisfy the Pauli exclusion principle, which states that a maximum of two electrons can occupy any given orbital and then only if the spins of the electrons are opposed. Equation (2.42) is often abbreviated as the product of the diagonal elements of the matrix,

$$\Psi = \frac{1}{\sqrt{2n!}} |\varphi_1(1)\bar{\varphi}_1(2) \ldots \varphi_n(2n-1)\bar{\varphi}(2n)|, \qquad (2.43)$$

where the bar over φ_i means a β spin and no bar means an α spin.

The molecular orbitals, φ_i, are approximated by linear combinations of AOs, χ_j, for all the atoms in a molecule as expressed by equation (2.9).

2.B.2 Energy Corresponding to Ψ

The expectation value of the energy E can be calculated using equations (2.1) and (2.4):

$$E = \int \Psi^* \mathcal{H} \Psi \, dv$$

$$= \int \frac{1}{\sqrt{2n!}} \left|\varphi_1(1)\bar{\varphi}_1(2) \ldots \varphi_n(2n)\right|^* \left\{ \sum_{i=1}^{2n} \mathcal{H}_{core}(i) + \sum_{i>j}^{2n} \frac{e^2}{r_{ij}} \right\}$$

$$\times \frac{1}{\sqrt{2n!}} |\varphi_1(1)\bar{\varphi}_1(2) \ldots \bar{\varphi}_n(2n)| \, dv, \qquad (2.44)$$

where

$$\mathcal{H}_{core}(i) = -\frac{\hbar^2}{2m} \sum_i^{2n} \nabla^2 - \sum_{i>k}^{2n} \frac{Z_k e^2}{r_{ik}}. \qquad (2.45)$$

Equation (2.44) has the form

$$E = 2 \sum_i^n I_i + \sum_{i,j} (2J_{ij} - K_{ij}), \qquad (2.46)$$

where I_i is the core energy for electron i. For example, I_1 is expressed as

$$I_1 = \int \varphi_1^*(1)\bar{\varphi}_1^*(2) \ldots \bar{\varphi}_n^*(2n) \mathcal{H}_{core}(1)\varphi_1(1)\bar{\varphi}_1(2) \ldots \bar{\varphi}_n(2n) \, dv(1) \ldots dv(2n)$$

$$= \int \varphi_1^*(1)\mathcal{H}_{core}(1)\varphi_1(1) \, dv(1) \int \bar{\varphi}_1^*(2)\bar{\varphi}_1^*(2) \, dv(2) \ldots \int \bar{\varphi}_n^*(2n)\bar{\varphi}_n^*(2n) \, dv(2n)$$

$$= \int \varphi_1^*(1)\mathcal{H}_{core}(1)\varphi_1(1) \, dv(1). \qquad (2.47)$$

Thus, I_i can be defined by

$$I_i = \int \varphi_i^*(1)\mathcal{H}_{core}(1)\varphi_i(1) \, dv(1). \qquad (2.48)$$

J_{ij} in equation (2.46) is the Coulomb integral and is defined by

$$J_{ij} = \int \varphi_i^*(1)\varphi_i^*(1)\frac{e^2}{r_{12}}\varphi_j(2)\varphi_j(2)\,dv(1)\,dv(2). \tag{2.49}$$

Finally, K_{ij} is the exchange integral which is defined by

$$K_{ij} = \int \varphi_i^*(1)\varphi_j^*(1)\frac{e^2}{r_{12}}\varphi_j(2)\varphi_i(2)\,dv(1)\,dv(2). \tag{2.50}$$

2.B.3 LCAO Self-consistent Field (SCF) Equations

The idea of a SCF is coupled with the variation method. The energy is minimized with respect to the adjustable parameters contained in the wavefunctions. This procedure was developed by Roothaan[3] using the Hartree–Fock method in which the basic equations for the SCF approach, for atomic systems, were laid down by Hartree and the wavefunctions were given in the form of a Slater determinant by Fock. The Hartree–Fock equations, as presented by Roothaan, are expressed as

$$(\mathbf{F} - \varepsilon_i \mathbf{S})\mathbf{C}_i = 0, \tag{2.51}$$

where the matrices $\mathbf{F}$, $\mathbf{S}$ and $\mathbf{C}_i$ have the following elements:

$$F_{rs} = H_{rs} + \sum_{j,t,u} C_{tj} C_{uj}[2\langle rs|tu\rangle - \langle rt|su\rangle], \tag{2.52}$$

$$S_{rs} = \int \chi_r^* \chi_s \, dv, \tag{2.53}$$

where

$$\mathbf{C}_i = (C_{1i}\ C_{2i}\ \ldots\ C_{mi}), \tag{2.54}$$

$$H_{rs} = \int \chi_r^* \mathcal{H}_{\text{core}} \chi_s \, dv, \tag{2.55}$$

$$\langle rs|tu\rangle = \int \chi_r^*(1)\chi_s^*(1)\frac{e^2}{r_{12}}\chi_t(2)\chi_u(2)\,dv(1)\,dv(2). \tag{2.56}$$

If the neglect of differential overlap approximation is introduced, whereby electron repulsion integrals involving the overlap distributions are assumed to be negligibly small, the above matrix elements become

$$F_{rr} = H_{rr} + \tfrac{1}{2}P_{rr}\langle rr|rr\rangle + \sum_{t\neq r} P_{tt}\langle rr|tt\rangle, \tag{2.57}$$

$$F_{rs} = H_{rs} - \tfrac{1}{2}P_{rs}\langle rr|ss\rangle, \tag{2.58}$$

where $r \neq s$ and

$$P_{rs} = 2\sum_{j}^{n} C_{rj}C_{sj} \tag{2.59}$$

Then, ε_i are the roots of the following determinantal equation:

$$\begin{vmatrix} F_{11} - \varepsilon S_{11} & F_{12} - \varepsilon S_{12} & \cdots & F_{1m} - \varepsilon S_{1m} \\ F_{21} - \varepsilon S_{21} & F_{22} - \varepsilon S_{22} & \cdots & F_{2m} - \varepsilon S_{2m} \\ \vdots & \vdots & & \vdots \\ F_{m1} - \varepsilon S_{m1} & F_{m2} - \varepsilon S_{m2} & \cdots & F_{mm} - \varepsilon S_{mm} \end{vmatrix} = 0. \tag{2.60}$$

2.B.4 . Unrestricted SCF Method

When the total number of α spins is different from the total number of β spins, the difference between the α and β spins is distinguishable. In such a case, the wavefunction is described as an unrestricted single determinant. Then, the function Ψ for $p + q$ electrons, which consists of p α-spin electrons and q β-spin electrons, is expressed as

$$\Psi = |\varphi_1^\alpha(1)\varphi_1^\beta(1)\ldots\varphi_p^\alpha(p)\varphi_p^\beta(p)\ldots\varphi_q^\alpha(q)\varphi_q^\beta(q)|, \tag{2.61}$$

where a normalizing factor is omitted. The MOs for the α and β spins can be written as LCAOs:

$$\varphi_i^\alpha = \sum_r C_{ri}^\alpha \chi_r, \tag{2.62}$$

$$\varphi_i^\beta = \sum_r C_{ri}^\beta \chi_r. \tag{2.63}$$

The SCF equations which satisfy the coefficients C_{ri}^α and C_{ri}^β, and the MO energies ε_i^α and ε_i^β, are expressed by

$$(\mathbf{F}^\alpha - \varepsilon_i^\alpha \mathbf{S})\mathbf{C}_i^\alpha = 0, \tag{2.64}$$

$$(\mathbf{F}^\beta - \varepsilon_i^\beta \mathbf{S})\mathbf{C}_i^\beta = 0. \tag{2.65}$$

Thus, the matrix elements are given, within the neglect of differential overlap approximation, as

$$F_{rr}^\alpha = H_{rr} + P_{rr}^\beta \langle rr|rr \rangle + \sum_{s \neq r} P_{rs} \langle rr|ss \rangle, \tag{2.66}$$

$$F_{rs}^\alpha = H_{rs} - P_{rs}^\alpha \langle rr|ss \rangle, \tag{2.67}$$

$$F_{rr}^\beta = H_{rr} + P_{rr}^\alpha \langle rr|rr \rangle + \sum_{s \neq r} P_{rs} \langle rr|ss \rangle, \tag{2.68}$$

$$F_{rs}^\beta = H_{rs} - P_{rs}^\beta \langle rr|ss \rangle, \tag{2.69}$$

where

$$P_{rs}^{\alpha} = \sum_{i}^{\text{occ.}} C_{ri}^{\alpha} C_{si}^{\alpha}, \qquad (2.70)$$

$$P_{rs}^{\beta} = \sum_{i}^{\text{occ.}} C_{ri}^{\beta} C_{si}^{\beta}, \qquad (2.71)$$

$$P_{rs} = P_{rs}^{\alpha} + P_{rs}^{\beta}. \qquad (2.72)$$

2.B.5 Empirical MO Methods

As examples of empirical MO procedures we introduce the Hückel[7] and extended Hückel[8] methods. These incorporate the most simple approximations and the resulting Hamiltonian is expressed in terms of a series of one-electron expressions, $\mathcal{H}_{\text{eff}}$:

$$\mathcal{H} = \sum_{i=1}^{2n} \mathcal{H}_{\text{eff}}(i). \qquad (2.73)$$

This approximation gives a qualitative account containing the substantial features of the electronic structure and electronic states of molecules in spite of the simplified model. The equation which we should solve by applying the variation method is given as

$$(\mathcal{H} - \varepsilon_i \mathbf{S})\mathbf{C}_i = 0, \qquad (2.74)$$

where the elements of the matrices $\mathcal{H}$ and $\mathbf{S}$ are

$$\mathcal{H}_{rs} = \int \chi_r^* \mathcal{H} \chi_s \, dv, \qquad (2.75)$$

$$S_{rs} = \int \chi_r^* \chi_s \, dv \qquad (2.76)$$

and $\mathbf{C}_i$ is a column vector of the LCAO coefficients, C_m, given by equation (2.9). Substituting equations (2.73), (2.75) and (2.76) into equation (2.74), we have the following equation:

$$\sum_{r,s} C_{ri}^* C_{si} H_{rs} - \varepsilon_i \sum_{r,s} C_{ri}^* C_{si} S_{rs} = 0, \qquad (2.77)$$

$$\varepsilon_i = \sum_{r,s} C_{ri}^* C_{si} H_{rs}. \qquad (2.78)$$

Thus, the total electronic energy is given as

$$E = 2 \sum_{i}^{n} \varepsilon_i. \qquad (2.79)$$

In the Hückel method, all the overlap integrals for the valence electrons, S_{rs}, are zero, even those between adjacent atoms, except for appropriate one-centre overlap integrals which are taken to be unity. The Coulomb integrals, H_{rr}, are given the same value for identical atoms and all the exchange integrals, H_{rs}, between adjacent atoms are assumed to be equal whereas those between the other atoms are set to zero. In empirical MO methods these integrals are determined on the basis of experimental data. As an example, in the extended Hückel method the appropriate integrals are given by

$$H_{rr} = -I_r, \qquad (2.80)$$

$$H_{rs} = 1.75 S_{rs} \left(\frac{H_{rr} + H_{ss}}{2} \right), \qquad (2.81)$$

where I_r is the ionization potential for electrons in the AO, r.

2.B.6 Semi-empirical MO Methods

Today, we have several semi-empirical MO methods to choose from. Presented here are some methods which are very popular and are readily available as "black box" computer programs. Thus, we introduce the CNDO/2, INDO, MINDO/2 and MINDO/3 methods as being amongst the most commonly employed semi-empirical MO procedures.

(a) *CNDO/2 Method*

The complete neglect of differential overlap (CNDO) approximation for electron repulsion integrals was introduced by Pople, Santry and Segal.[9,10] The first paper related to the CNDO method was later slightly modified to produce the CNDO/2 method.[11]

Let us introduce the CNDO/2 method briefly. In this approach, the following basic approximations are encountered.

Approximation 1 The two electron integrals in equation (2.56) are taken to be zero except for $r = s$ and $t = u$ and the matrix elements of the SCF equations are given by equations (2.57) and (2.58).

Approximation 2 If AO r and AO s are associated with atoms A and B, respectively, the two electron integrals are given as

$$\langle rr|rr \rangle = \gamma_{AA} \qquad (2.82)$$

$$\langle rr/ss \rangle = \langle 2s_A 2s_A | 2s_B 2s_B \rangle = \gamma_{AB}, \qquad (2.83)$$

where $2s_A$ means a 2s orbital on atom A, and r and s represent 2s, $2p_x$, $2p_y$ or $2p_z$ for first row elements.

Approximation 3 The diagonal elements involving the core integrals are approximated as

$$H_{rr} = -\tfrac{1}{2}(I_r + A_r) - (Z_A - \tfrac{1}{2})\gamma_{AA} - \sum_{B \neq A} Z_B \gamma_{AB}, \qquad (2.84)$$

$$H_{rs} = \tfrac{1}{2}S_{rs}(\beta_A + \beta_B), \qquad (2.85)$$

where I_r and A_r are respectively the ionization potential of the appropriate average atomic state and the atomic electron affinity. β_A is the bonding parameter, and depends only on the nature of the atom A. The values of parameters normally used in the CNDO/2 method are summarized in Table 2.1.

Table 2.1 Parameters used in the CNDO/2 method

	H	Li	Be	B	C	N	O	F
$(I_s + A_s)/2$	7.176	3.106	5.946	9.594	14.051	19.316	25.390	32.272
$(I_p + A_p)/2$		1.258	2.563	4.001	5.572	7.275	9.111	11.080
β_A^0	9	9	13	17	21	25	31	39

Energies are expressed in electron-volts.

Pople and Beveridge[6] have calculated conformational energies, dipole moments and force constants of a large number of molecules with this MO method and have obtained satisfactory agreement between the calculated and observed results. As predicted from the parameterization, calculated values of the ionization potential and excitation energy are somewhat higher than the observed ones, but it is suggested that these deviations may be modified by a correction of the parameters used in the calculations.

(b) *INDO Method*

The approximations which are incorporated in this method are almost the same as in the CNDO/2 procedure except for the estimation of the two-electron integrals. One-centre two-electron exchange integrals are retained in the INDO method but are neglected in the CNDO method. In equation (2.56) the following integrals are taken into account in the INDO approach:

$$\langle ss|ss \rangle = \langle ss|xx \rangle = F^0 = \gamma_{AA}, \qquad (2.86)$$

$$\langle xx|xx \rangle = F^0 + \tfrac{4}{25}F^2, \qquad (2.87)$$

$$\langle xx|yy \rangle = F^0 - \tfrac{2}{25}F^2, \qquad (2.88)$$

$$\langle sx|sx \rangle = \tfrac{1}{3}G^1, \qquad (2.89)$$

$$\langle xy|xy \rangle = \tfrac{3}{25}F^2, \qquad (2.90)$$

where s refers to a 2s AO and x and y are the $2p_x$ and $2p_y$ orbitals, respectively. The values of F^0 are calculated using the Slater AOs and the values of the Slater–Condon repulsion integrals F^2 and G^1 are used as semi-empirical parameters as shown in Table 2.2. For this reason, this is referred to as the intermediate neglect of differential overlap (INDO) method.[12]

Table 2.2 Values of the Slater–Condon repulsion integrals, F^2 and G^1, for some atoms

Atom	G^1	F^2
Li	0.092 012	0.049 865
Be	0.140 7	0.089 125
B	0.199 265	0.130 41
C	0.267 708	0.173 72
N	0.346 029	0.219 055
O	0.434 23	0.266 415
F	0.532 305	0.315 80

Energies are expressed in electron-volts.

The trends in the values of many molecular properties calculated by the CNDO/2 and INDO methods are in general similar, as is to be expected from the basic approximations involved. For spin–spin couplings INDO is much better than CNDO/2, as it is also for hyperfine couplings.

(c) *MINDO/2 and MINDO/3 Methods*

Dewar et al.[13-15] have developed a MO method which is aimed at reproducing experimental results rather than *ab initio* results, which is the main goal of the CNDO and INDO procedures. Several parameters are introduced in the estimation of the necessary integrals within the INDO method and improved results are obtained by modified INDO methods for heats of formation, force constants, ionization potentials and atomic distances in molecules. Thus, the method is referred to as the modified INDO (MINDO) method.

In the MINDO/2 procedure,[14] the two-centre, two-electron integrals are estimated using Ono's approximation:

$$\gamma_{AB} = \frac{14.99}{\sqrt{R_{AB}^2 + 51.8318\left(\frac{1}{F_A^0} + \frac{1}{F_B^0}\right)^2}} \qquad (2.91)$$

where R_{AB} is the distance between atoms A and B. The exchange integrals,

H_{rs}, are estimated as

$$H_{rs} = \beta_{AB} S_{rs}(I_r + I_s), \qquad (2.92)$$

where r and s represent 2s or 2p orbitals, and I is the ionization potential.

$$I_s = U_s + (Z_A - 1)(F_A^0 - \tfrac{1}{6}G_A^1), \qquad (2.93)$$

$$I_p = U_p + \tfrac{1}{3}(Z_A - 1)(3F_A^0 - \tfrac{1}{6}G_A^1 - 0.28 F_A^2), \qquad (2.94)$$

where values for β_{AB}, U, F and G are found empirically. In the MINDO/2 procedure, to avoid overestimation for the net Coulomb repulsions between atoms, the core Hamiltonian, CH_{AB}, is expressed by

$$CH_{AB} = Z_A Z_B \gamma_{AB} + \left(\frac{Z_A Z_B}{R_{AB}} - Z_A Z_B \gamma_{AB}\right) \exp(-D_{AB} R_{AB}), \qquad (2.95)$$

where D_{AB} is a parameter characteristic of the atoms (A and B) and is found empirically.

Recently, this has been further modified by using adjustable parameters to give MINDO/3,[15] which gives better results for ground state molecular properties than the MINDO/2 method. In this method, one of the modifications is that, in the case of HN and HO bonds only, $\exp(-D_{AB} R_{AB})$ of equation (2.95) is replaced by $D_{AB} \exp(-R_{AB})$. However, in general the parameterization schemes used in the various MINDO procedures are more suitable to the estimation of ground state rather than the excited state properties of molecules. It is the latter which are important for NMR parameters.

2.B.7 Non-empirical Methods

Recently, non-empirical (*ab initio*) MO calculations for large molecules have become possible due to remarkable advances in computer technology. In these MO procedures the parameters which are handled in semi-empirical MO methods are not explicitly involved,[16] but the corresponding integrals are calculated.

After we assume equation (2.42) as the wavefunction, its best form is determined by using the variation method without further approximations. In the absence of applied fields the orbital energies depend only upon the radial components of the AOs. This comprises the Hartree–Fock wavefunction. Roothaan has developed as a basis set the radial functions $\varphi(r)$ which we need to determine. At present, Slater-type orbitals (STO) and Gaussian-type orbitals (GTO) are widely used. As the simplest example, we can write

STOs and GTOs in the following forms:

$$\varphi_{STO} \propto \exp(-\zeta r) \quad \text{(STO)}, \tag{2.96}$$

$$\varphi_{GTO} \propto \exp(-\alpha r^2) \quad \text{(GTO)}, \tag{2.97}$$

where ζ and α are treated as variational parameters.

In the LCAO-SCF method, using a minimal basis set of STOs, which has been used extensively, only 1s, 2s and 2p atomic orbitals are employed and are expressed as

$$\varphi_{1s} = (\zeta_1^3/\pi)^{1/2} \exp(-\zeta_1 r), \tag{2.98}$$

$$\varphi_{2s} = (\zeta_2^5/\pi)^{1/2} r \exp(-\zeta_2 r), \tag{2.99}$$

$$\varphi_{2p_x} = (\zeta_2^5/\pi)^{1/2} x \exp(-\zeta_2 r), \tag{2.100}$$

$$\varphi_{2p_y} = (\zeta_2^5/\pi)^{1/2} y \exp(-\zeta_2 r), \tag{2.101}$$

$$\varphi_{2p_z} = (\zeta_2^5/\pi)^{1/2} z \exp(-\zeta_2 r). \tag{2.102}$$

Calculations using STOs have been carried out for relatively small molecules. The calculations are time consuming largely because of the evaluation of a large number of two-electron integrals. In *ab initio* studies using GTOs the number of two-electron integrals to be evaluated is proportional to n^4, where n is the size of the basis set. In comparable semi-empirical calculations the number of similar integrals is proportional to n^2.

In order to reduce the computing time required, each STO is replaced by a linear combination of a small number of GTOs because integrals involving GTOs can be evaluated analytically. This technique was first used by Foster and Boys[17] and was later extensively developed by Pople *et al.*[18-21]

We will introduce briefly the STO-KG and STO 4-31 G sets.

(a) *STO-KG*

Each STO for 1s, 2s and 2p is replaced by a linear combination of K GTOs as

$$\varphi_{1s} = \sum_{k=1}^{K} d_{1s,k} g_{1s}(\alpha_{1k}, r), \tag{2.103}$$

$$\varphi_{2p_x} = \sum_{k=1}^{K} d_{2p_x,k} g_{2p_x}(\alpha_{2k}, r), \tag{2.104}$$

$$\varphi_{2p_y} = \sum_{k=1}^{K} d_{2p_y,k} g_{2p_y}(\alpha_{2k}, r), \tag{2.105}$$

$$\varphi_{2p_z} = \sum_{k=1}^{K} d_{2p_z,k} g_{2p_z}(\alpha_{2k}, r), \tag{2.106}$$

where

$$g_{1s}(\alpha_{1k}, r) = (2\alpha_{1k}/\pi)^{3/4} \exp(-\alpha_{1k}r^2), \quad (2.107)$$

$$g_{2p_x}(\alpha_{2k}, r) = (128\alpha_{2k}^5/\pi^3)^{1/4} x \exp(-\alpha_{2k}r^2), \quad (2.108)$$

$$g_{2p_y}(\alpha_{2k}, r) = (128\alpha_{2k}^5/\pi^3)^{1/4} y \exp(-\alpha_{2k}r^2), \quad (2.109)$$

$$g_{2p_z}(\alpha_{2k}, r) = (128\alpha_{2k}^5/\pi^3)^{1/4} z \exp(-\alpha_{2k}r^2). \quad (2.110)$$

The 2s STO is replaced by a linear combination of 1s GTOs. The constants d and α in equations (2.103)–(2.110) are variational parameters which have been determined by Pople et al.[18-21] Here, when $K = 3$, the function is referred to as STO-3G.

(b) *STO4-31G*

Pople et al.[18-21] have developed an extended basis set of atomic functions expressed as linear combinations of GTOs. Each inner shell is represented by a single basis function taken as a sum of four GTOs and each valence orbital is split into inner and outer parts described by three and one GTOs, respectively. The resulting function is referred to as a STO of type 4-31 G. For example, we can write atomic orbitals for hydrogen and carbon atoms. For the hydrogen atom,

$$\varphi'_{1s} = \sum_{k=1}^{4} d'_k g_s(\alpha'_k, r), \quad (2.111)$$

$$\varphi''_{1s} = d''_1 g_s(\alpha''_1, r); \quad (2.112)$$

and for the carbon atom,

$$\varphi'_{1s} = \sum_{k=1}^{4} d'_{1s,k} g_s(\alpha'_{1k}, r), \quad (2.113)$$

$$\varphi'_{2s} = \sum_{k=1}^{3} d'_{2s,k} g_s(\alpha'_{2k}, r), \quad (2.114)$$

$$\varphi'_{2p_x} = \sum_{k=1}^{3} d'_{2p,k} g_{p_x}(\alpha'_{2k}, r), \quad (2.115)$$

$$\varphi''_{2s} = d''_{2s,1} g_s(\alpha''_{21}, r), \quad (2.116)$$

$$\varphi''_{2p_x} = d''_{2p,1} g_{px}(\alpha''_{21}, r); \quad (2.117)$$

where φ' and φ'' represent the inner and outer parts of the valence shell, respectively.

Pople et al. have drawn the following general conclusion on *ab initio* MO calculations using Gaussian functions: the least-squares representation of STOs by a small sum of GTOs provides a rapidly convergent method for obtaining self-consistent MOs simulating those directly based on an STO set and the calculations provide reasonable results for molecular geometries and NMR parameters.

2.C VALENCE BOND METHOD

The valence bond (VB) method was initiated by Heitler and London[22] who used it in an attempt to explain the nature of the chemical bond in the hydrogen molecule. We will introduce this method by treating the six π electrons in the benzene molecule.

We first consider how to make electron pair bonds among the six π electrons in benzene. The scheme of bonding is shown below, in this scheme 1, 2, 3, 4, 5 and 6 represent the cores of the six carbon atoms. There are five bonding schemes

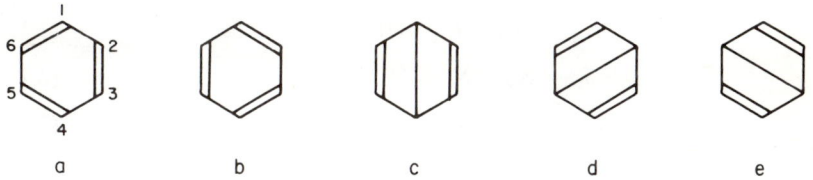

of which two are Kekulé structures (**a**, **b**) and the remaining three are Dewar structures (**c, d, e**). The total wave function of the molecule is constructed as a linear combination of these functions, which the molecule would possess if it could be represented by the structures **a, b, c, d** and **e**.

$$\Psi = C_a\Psi_a + C_b\Psi_b + C_c\Psi_c + C_d\Psi_d + C_e\Psi_e. \quad (2.118)$$

As a first step, each of the functions $\Psi_a, \Psi_b, \Psi_c, \Psi_d$ and Ψ_e must be determined. The Pauli exclusion principle implies that the probability of finding two electrons with the same spin in the same volume element is zero. For example, in structure **a**, where the π orbital functions on the carbon atoms are defined as $\varphi_1, \varphi_2, \varphi_3, \varphi_4, \varphi_5$ and φ_6, Ψ_a is given by

$$\Psi_a = \sqrt{\frac{1}{2^3}} \sum_R (-1)^R R \left[\frac{1}{\sqrt{6!}} \sum_P (-1)^P P \{\varphi_1(1)\alpha(1)\varphi_2(2)\beta(2) \right.$$

$$\left. \times \varphi_3(3)\alpha(3)\varphi_4(4)\beta(4)\varphi_5(5)\alpha(5)\varphi_6(6)\beta(6)\} \right], \quad (2.119)$$

where $1/\sqrt{2^3}$ is the normalization factor due to spin and R is a spin permutation operator. The multiplier P is an electron permutation operator; $(-1)^R$ and $(-1)^P$ are $+1$ for an even permutation and -1 for an odd permutation of spins.

The matrix components required for setting up the secular equation may be obtained from

$$H_{pq} = \int \Psi_p \mathcal{H} \Psi_q \, dv = \frac{1}{2^m}(kQ + l\alpha), \qquad (2.120)$$

where kQ represents the Coulomb integral Q multiplied by a coefficient k and α is the exchange integral, which is multiplied by a coefficient l, m is the number of bonds involved and $1/2^m$ is a normalization factor. The coefficients k and l are simply determined by Pauling's island method.[23] Thus, the matrix components are evaluated as follows:

$$H_{aa} = H_{bb} = Q + 1.5\alpha, \qquad (2.121)$$

$$H_{cc} = H_{dd} = H_{ee} = Q, \qquad (2.122)$$

$$H_{ab} = H_{cd} = H_{ce} = H_{ed} = \tfrac{1}{4}Q + 1.5\alpha, \qquad (2.123)$$

$$S_{aa} = S_{bb} = S_{cc} = S_{dd} = S_{ee} = 1, \qquad (2.124)$$

$$S_{ab} = S_{cd} = S_{ce} = S_{ed} = \tfrac{1}{4}, \qquad (2.125)$$

where the subscripts indicate the structures **a** to **e**, and S_{XY} is the overlap integral between the structures X and Y. If these matrix components are substituted into the secular equations, we can obtain the energies and corresponding eigenfunctions. As with the MO procedure in its simplest form, the valence bond method is also treated empirically. Values of Q and α are not usually computed but are obtained from comparison with experimental quantities.

2.D APPLICATIONS OF QUANTUM CHEMISTRY TO SECOND-ORDER MOLECULAR PROPERTIES

Second-order molecular properties such as nuclear shielding and spin–spin coupling depend upon the distortion of electron clouds by additional external perturbations; that is, in an NMR experiment they depend upon the electronic motion induced by an applied magnetic field. If, furthermore, the magnetic coupling of electrons to nuclei arises from the magnetic fields originating from the currents induced by electronic motion, these fields are responsible for the so-called nuclear shielding, and the nuclear spin–spin coupling. Theories for such second-order molecular properties require a study of the change in the molecular eigenfunctions, which may be found by using a perturbation method to describe the effects occurring when a magnetic field is applied. We shall consider these on the basis of the perturbation theory which was applied by Ramsey[24,25] to NMR parameters.

2.D.1 Nuclear Shielding[24]

In the presence of a magnetic field **B**, the motion of electrons can be described by replacing **p** $(=(\hbar/i)\nabla)$ in equation (2.4) with $\mathbf{p} + e\mathbf{A}$, where **p** is the generalized momentum and **A** is the vector potential due to the applied magnetic field. **A** is defined as

$$\mathbf{A} = \tfrac{1}{2}\mathbf{B} \times \mathbf{r}, \qquad \text{div } \mathbf{A} = 0, \qquad \text{curl } \mathbf{A} = \mathbf{B}. \tag{2.126}$$

The Hamiltonian of the electronic system of any molecule considered, in the presence of a magnetic field, is given by

$$\mathcal{H} = \frac{1}{2m} \sum_j (\mathbf{p}_j + e\mathbf{A}_j)^2 + V$$

$$= \frac{1}{2m} \sum_j \left(\frac{\hbar}{i}\nabla_j + e\mathbf{A}_j\right)^2 + V, \tag{2.127}$$

where V is the electrostatic potential energy term. The first term in equation (2.127) may be written as

$$(\mathbf{p}_j + e\mathbf{A}_j)^2 = -\hbar^2 \nabla_j^2 - i\hbar \nabla_j \cdot \mathbf{A}_j - i\hbar e \mathbf{A}_j \cdot \nabla_j + e^2 \mathbf{A}_j^2$$

$$= -\hbar^2 \nabla_j^2 - i\hbar e \text{ div } \mathbf{A}_j - i\hbar e \mathbf{A}_j \cdot \nabla_j + e^2 \mathbf{A}_j^2. \tag{2.128}$$

We assume that the applied magnetic field lies in the $+z$ direction, and the origin of the coordinates is chosen at the nucleus for which the shielding is desired. We have

$$\mathbf{A}_j = \tfrac{1}{2}\mathbf{B} \times \mathbf{r}_j + \left(\frac{\mu_0}{4\pi}\right)\frac{\boldsymbol{\mu} \times \mathbf{r}_j}{|\mathbf{r}_j|^3}, \qquad \mathbf{B} = (0\ 0\ B). \tag{2.129}$$

where **μ** is the magnetic moment of the nucleus concerned. The vector $\mathbf{r}_j$ is the radius vector from the origin, at the nucleus, to the position of electron j,

$$\mathbf{r}_j = \mathbf{i}x_j + \mathbf{j}y_j + \mathbf{k}z_j, \tag{2.130}$$

where **i, j** and **k** are unit vectors fixed in the laboratory. Thus, the components of $\mathbf{A}_j$ are obtained as

$$
\begin{aligned}
\mathbf{A}_{x_j} &= -\tfrac{1}{2} B y_i - \left(\frac{\mu_0}{4\pi}\right) \frac{\mu y_j}{|\mathbf{r}_j|^3}, \\
\mathbf{A}_{y_j} &= \tfrac{1}{2} B x_i + \left(\frac{\mu_0}{4\pi}\right) \frac{\mu x_j}{|\mathbf{r}_j|^3}, \\
\mathbf{A}_{z_j} &= 0.
\end{aligned}
\qquad (2.131)
$$

Substituting equations (2.129) and (2.131) into equation (2.128), we have

$$
\begin{aligned}
(\mathbf{p}_j + e\mathbf{A}_j)^2 &= -\hbar^2 \mathbf{V}_j^2 - i\hbar \left(\tfrac{1}{2} B + \left(\frac{\mu_0}{4\pi}\right) \frac{\mu}{|\mathbf{r}_j|^3}\right) \\
&\quad \times \left(-y_j \frac{\partial}{\partial x_j} + x_j \frac{\partial}{\partial y_j}\right) + e^2 \left(\tfrac{1}{2} B + \left(\frac{\mu_0}{4\pi}\right) \frac{\mu}{|\mathbf{r}_j|^3}\right) \\
&\quad \times (x_j^2 + y_j^2).
\end{aligned}
\qquad (2.132)
$$

Therefore, we obtain the Hamiltonian corresponding to equation (2.127):

$$
\begin{aligned}
\mathcal{H} &= \frac{1}{2m} \sum_j (\hbar^2 \mathbf{V}_j^2 + V) - \sum_j \left(B + \left(\frac{\mu_0}{4\pi}\right) \frac{2\mu}{|\mathbf{r}_j|^3}\right) m_{zj}^0 \\
&\quad + \sum_j \frac{e^2}{8m} \left(B + \left(\frac{\mu_0}{4\pi}\right) \frac{2\mu}{|\mathbf{r}_j|^3}\right)^2 (x_j^2 + y_j^2),
\end{aligned}
\qquad (2.133)
$$

where

$$
m_{zj}^0 = -\frac{e\hbar}{2im} \left(x_j \frac{\partial}{\partial y_j} - y_i \frac{\partial}{\partial x_j}\right). \qquad (2.134)
$$

Now, $\mathcal{H}$ may be written as

$$
\mathcal{H} = \mathcal{H}_0 + \mathcal{H}^1 + \mathcal{H}^2. \qquad (2.135)
$$

Here $\mathcal{H}_0$, $\mathcal{H}^1$ and $\mathcal{H}^2$ correspond to the first, second and third terms in equation (2.133), respectively. We treat $\mathcal{H}^1 + \mathcal{H}^2$ as a perturbation on the eigenfunctions and eigenvalues of $\mathcal{H}_0$. The first-order and second-order perturbation energies are obtained by using the perturbation method presented in Section 2.A.3 as

$$
E_k' = \langle \Psi_k^0 | \mathcal{H}^1 + \mathcal{H}^2 | \Psi_k^0 \rangle, \qquad (2.136)
$$

$$
E_k'' = \sum_{m \neq k} \frac{\langle \Psi_k^0 | \mathcal{H}^1 + \mathcal{H}^2 | \Psi_m^0 \rangle \langle \Psi_m^0 | \mathcal{H}^1 + \mathcal{H}^2 | \Psi_k^0 \rangle}{E_m - E_k}. \qquad (2.137)
$$

The electronic energy, E_k, of the molecule concerned, in the presence of the perturbations due to μ and B, can be calculated by collecting all terms whose dependence on μ and B is given by the linear product, μB. Thus, the nuclear shielding tensor σ has components which are obtained from the energy relationship

$$E_k = \sigma \mu B. \tag{2.138}$$

By selecting the terms linear in μB from equations (2.136) and (2.137), we obtain

$$E'_k = \frac{e^2}{8m} \langle \Psi^0_k | \sum_j \frac{4\mu B}{|\mathbf{r}_j|^3} (x_j^2 + y_j^2) | \Psi^0_k \rangle$$

$$= \frac{e^2}{2m} \mu B \langle \Psi^0_k | \sum_j \frac{x_j^2 + y_j^2}{|\mathbf{r}_j|^3} | \Psi^0_k \rangle \tag{2.139}$$

and

$$E''_k = \sum_{m \neq k} \frac{2\mu B}{E_m - E_k} \Big\{ \langle \Psi^0_k | - \sum_j m^0_{zj} | \Psi^0_m \rangle$$

$$\times \langle \Psi^0_m | - \sum_j \frac{m^0_{zj}}{|\mathbf{r}_j|^3} | \Psi^0_k \rangle + \langle \Psi^0_k | - \sum_j \frac{m^0_{zj}}{|\mathbf{r}_j|^3} | \Psi^0_m \rangle$$

$$\times \langle \Psi^0_m | - \sum_j m^0_{zj} | \Psi^0_k \rangle. \tag{2.140}$$

Thus we have

$$E = E' + E''. \tag{2.141}$$

The nuclear shielding may be evaluated from equations (2.138)–(2.141), as detailed in Chapter 3.

2.D.2 Nuclear Spin–Spin Coupling[25]

The general equations for electron spin and orbital contributions to nuclear spin–spin couplings are derived using perturbation methods. The corresponding Hamiltonians are given by

$$\mathcal{H}_0 = \frac{1}{2m} \sum_j \left(\frac{\hbar}{i} \right) \nabla_j^2 + \bar{V} + \mathcal{H}_{LL} + \mathcal{H}_{LS} + \mathcal{H}_{SS}, \tag{2.142}$$

$$\mathcal{H}^1 = \frac{e\hbar \mu_B}{i} \sum_{kN} \frac{\gamma_N \mathbf{I}_N \times \mathbf{r}_{kN}}{|\mathbf{r}_{kN}|^3} \nabla_k + \frac{e^2 \hbar^2 \mu_B^2}{2m} \sum_{kNN'} \gamma_N \gamma_{N'}$$

$$\times \left(\mathbf{I}_N \times \frac{\mathbf{r}_{kN}}{|\mathbf{r}_{kN}|^3} \right) \left(\mathbf{I}_{N'} \times \frac{\mathbf{r}_{kN'}}{|\mathbf{r}_{kN'}|^3} \right), \tag{2.143}$$

$$\mathcal{H}^2 = \frac{\mu_0 \mu_B \hbar}{2\pi} \sum_{jN} \gamma_N \left\{ \frac{3(\mathbf{S}_j \cdot \mathbf{r}_{jN})(\mathbf{I}_N \cdot \mathbf{I}_{jN})}{|\mathbf{r}_{jN}|^5} - \frac{(\mathbf{S}_j \cdot \mathbf{I}_N)}{|\mathbf{r}_{jN}|^3} \right\}, \qquad (2.144)$$

$$\mathcal{H}^3 = \frac{4\mu_0 \mu_B \hbar}{3} \sum_{jN} \gamma_N \, \delta(|\mathbf{r}_{jN}|)(\mathbf{S}_j \cdot \mathbf{I}_N), \qquad (2.145)$$

where V is the electrostatic interaction and $\mathcal{H}_{LL}$, $\mathcal{H}_{LS}$ and $\mathcal{H}_{SS}$ are magnetic interactions among the electrons. I represents the nuclear spin angular momentum, and S the electron spin angular momentum. γ is the gyromagnetic ratio, μ_0 the permittivity of free space, μ_B the Bohr magneton and δ the Dirac delta function. $\mathcal{H}_0$ gives the total electronic kinetic energy and $\mathcal{H}^1$ the magnetic interactions between electron orbital motions and nuclear magnetic moments. $\mathcal{H}^2$ gives the magnetic dipolar interactions between electrons and nuclear magnetic moments and $\mathcal{H}^3$ gives the contact interaction between electron spins and nuclear spins. The corresponding second-order perturbation energy is given by an equation similar to (2.137) in which $\mathcal{H}^2$ and $\mathcal{H}^3$ are replaced by equations (2.144) and (2.145).

The electron-coupled nuclear spin interaction energy is given as [25]

$$E_{NN'} = hJ_{NN'}\mathbf{I}_N \cdot \mathbf{I}_{N'} + h\mathbf{I}_N \cdot \mathbf{J}_{NN'} \cdot \mathbf{I}_{N'}, \qquad (2.146)$$

where $J_{NN'}$ represents the scalar spin–spin coupling constant when nuclear spins N and N' are coupled and $\mathbf{J}_{NN'}$ is the second rank coupling tensor, or dyadic, whose trace is zero. Molecular motions in the liquid state average the second term of equation (2.146) to zero. Through a comparison of the second-order perturbation energy and equation (2.146), the appropriate spin–spin coupling may be evaluated. The calculation of the NMR parameters σ and J is dealt with in greater detail in Chapters 3 and 4, respectively.

REFERENCES

1. H. Eyring, J. Walter and G. E. Kimball, "Quantum Chemistry", John Wiley & Sons, New York (1944).
2. H. F. Hameka, "Advanced Quantum Chemistry", Addison-Wesley, Boston (1965).
3. C. C. J. Roothaan, *Rev. mod. Phys.*, **32**, 179 (1960).
4. J. A. Pople, J. W. McIver and N. S. Ostlund, *J. Chem. Phys.*, **49**, 2960, 2965 (1968).
5. M. J. S. Dewar, "The Molecular Orbital Theory of Organic Chemistry", McGraw-Hill, New York (1969).
6. J. A. Pople and D. L. Beveridge, "Approximate Molecular Orbital Theory", McGraw-Hill, New York (1970).
7. E. Hückel, *Z. Phys.*, **70**, 204 (1931).

8. R. Hoffman, *J. Chem. Phys.*, **39**, 1397 (1963).
9. J. A. Pople, D. P. Santry and G. A. Segal, *J. Chem. Phys.*, **43**, S129 (1965).
10. J. A. Pople and G. A. Segal, *J. Chem. Phys.*, **43**, S136 (1965).
11. J. A. Pople and G. A. Segal, *J. Chem. Phys.*, **44**, 3289 (1966).
12. J. A. Pople, D. L. Beveridge and P. A. Dobosh, *J. Chem. Phys.*, **47**, 2026 (1967).
13. N. C. Baird and M. J. S. Dewar, *J. Chem. Phys.*, **50**, 1262 (1969).
14. M. J. S. Dewar and E. Haselbach, *J. Am. chem. Soc.*, **92**, 590 (1970).
15. R. C. Bingham, M. J. S. Dewar and D. H. Lo, *J. Am. chem. Soc.*, **97**, 1285 (1975).
16. D. B. Cook, "*Ab initio* Valence Calculations in Chemistry", Butterworth, London (1974).
17. J. M. Foster and S. F. Boys, *Rev. mod. Phys.*, **32**, 303 (1960).
18. W. J. Hehre, R. F. Stewart and J. A. Pople, *J. Chem. Phys.*, **51**, 2657 (1969).
19. W. J. Hehre, R. Ditchfield, R. F. Stewart and J. A. Pople, *J. Chem. Phys.*, **52**, 2769 (1970).
20. M. D. Newton, W. A. Lathan, W. J. Hehre and J. A. Pople, *J. Chem. Phys.*, **52**, 4064 (1970).
21. R. Ditchfield, W. J. Hehre and J. A. Pople, *J. Chem. Phys.*, **52**, 5001 (1970).
22. W. Heitler and F. London, *Z. Phys.*, **44**, 455 (1927).
23. L. Pauling, *J. Chem. Phys.*, **1**, 280 (1933).
24. N. S. Ramsey, *Phys. Rev.*, **78**, 699 (1950).
25. N. S. Ramsey, *Phys. Rev.*, **91**, 303 (1953).

3
Nuclear Shielding

After obtaining an NMR spectrum and analysing it in terms of the various parameters, an understanding is then required of the relationship between these parameters and various aspects of molecular electronic structure. Nuclear shielding is usually discussed in terms of the electronic distribution obtained from molecular orbital (MO) calculations.

In addition to such a general description it is necessary to consider the effects of minor influences for nuclei which have rather small chemical shift ranges. Such nuclei are those of hydrogen, where ring currents and local magnetic anisotropies may play an important role. Ring currents may be treated by classical or quantum mechanical procedures as discussed in Section 3.D. We begin with some general remarks on the theoretical background to nuclear shielding.

3.A GENERAL THEORETICAL BACKGROUND

The shielding tensor for a nucleus in an isolated molecule is usually expressed in terms of the theory presented by Ramsey.[1] The basis of this model involves placing the molecule of interest in a magnetic field B, which has a given orientation.† As discussed in Section 2.D, the molecular electronic energy is subject to perturbations due to interactions between the electrons, on one hand, and the external field, and a field due to the magnetic dipole μ_A, of the nucleus A concerned, on the other. The electronic eigenfunctions for the ground and excited states are obtained in the absence of magnetic fields. The resulting zero-field functions, Ψ^0, are then used to represent the ground state electronic structure in the presence of a magnetic field. In principle, a variational approach to the problem is also permissible. However, it is the perturbation procedure, adopted by Ramsey, which is the most widely employed.

† In this chapter B, rather than B_0, is used to signify the applied magnetic field in the NMR experiment due to the use of numerous subscripts.

The perturbation theory is developed by means of a Taylor series expansion of $\Psi(B, \mu_A)$ around zero-field values of B and μ_A, in which it is assumed that B and μ_A are small perturbations:

$$\Psi(B, \mu_A) = \Psi^0 + \Psi^{(1,0)}B + \sum_A \Psi_A^{(0,1)}\mu_A + \ldots, \qquad (3.1)$$

where $\Psi^{(0,1)}$, etc., are the expansion coefficients, the first post-superscript referring to B and the second to μ_A.

The corresponding Hamiltonian operator becomes

$$\mathcal{H}(B, \mu_A) = \mathcal{H}^0 + \mathcal{H}^{(1,0)} \cdot B + \tfrac{1}{2} B \cdot \mathcal{H}^{(2,0)} \cdot B$$
$$+ \sum_A \mu_A \cdot \mathcal{H}_A^{(0,1)} + \tfrac{1}{2} \sum_A \mu_A \cdot \mathcal{H}^{(1,1)} \cdot B + \ldots. \qquad (3.2)$$

The derivation of the various operator expressions in equation (3.2) is discussed in detail by Slichter.[2]

The eigenvalues obtained from the solution of the corresponding Schrödinger equation are written as

$$E(B, \mu_A) = E_0 - \sum_\alpha M_\alpha B_\alpha - \sum_\alpha \mu_{A\alpha} B_\alpha$$
$$- \tfrac{1}{2} \sum_\alpha \sum_\beta B_\alpha \chi_{\alpha\beta} B_\beta + \sum_\alpha \sum_\beta B_\alpha \sigma_{A\alpha\beta} \mu_{A\beta} \ldots, \qquad (3.3)$$

where α and β refer to Cartesian components, x, y or z; M is the permanent magnetic moment of the molecule, which is zero for closed-shell systems; the third term in equation (3.3) represents the direct interaction between the nuclear magnetic moment and the external magnetic field; the fourth term describes the molecular diamagnetic polarization; the final term relates to the nuclear shielding. Equation (3.3) may be alternatively written as

$$E(B, \mu_A) = E_0 + \sum_\alpha E_\alpha^{(1,0)} B_\alpha + \sum_\alpha E_{A\alpha}^{(0,1)} \mu_{A\alpha}$$
$$+ \tfrac{1}{2} \sum_\alpha \sum_\beta B_\alpha E_{\alpha\beta}^{(2,0)} B_\beta + \sum_\alpha \sum_\beta B_\alpha E_{A\alpha\beta}^{(1,1)} \mu_{A\beta} + \ldots. \qquad (3.4)$$

Thus, in order to obtain a value for $\sigma_{A\alpha\beta}$, it is apparent that $E_{A\alpha\beta}^{(1,1)}$ has to be determined as indicated by equation (2.138).

By means of Rayleigh–Schrödinger perturbation theory,[3] $E_{A\alpha\beta}^{(1,1)}$ is obtained as

$$E_{A\alpha\beta}^{(1,1)} = \langle \Psi^0 | \mathcal{H}_A^{(1,1)} | \Psi^0 \rangle + \langle \Psi^0 | \mathcal{H}_{A\alpha}^{(0,1)} | \Psi_\beta^{(1,0)} \rangle$$
$$+ \langle \Psi^0 | \mathcal{H}_\alpha^{(1,0)} | \Psi_{A\beta}^{(0,1)} \rangle. \qquad (3.5)$$

$\Psi_\beta^{(1,0)}$ and $\Psi_{A\beta}^{(0,1)}$ may be expanded in terms of the complete set of

3 NUCLEAR SHIELDING

unperturbed functions, Ψ_k. This gives

$$\Psi_\beta^{(1,0)} = \sum_{k \neq 0}^{\infty} C_{k\beta}^{(1,0)} \Psi_k, \tag{3.6}$$

where

$$C_{k\beta}^{(1,0)} = -(E_k - E_0)^{-1} \langle \Psi_k | \mathcal{H}_\beta^{(1,0)} | \Psi_0 \rangle. \tag{3.7}$$

The expression for $\sigma_{A\alpha\beta}$ becomes

$$\sigma_{A\alpha\beta} = \langle \Psi_0 | \mathcal{H}_{A\alpha\beta}^{(1,1)} | \Psi_0 \rangle - \sum_{k \neq 0}^{\infty} (E_k - E_0)^{-1}$$
$$\times [\langle \Psi_0 | \mathcal{H}_\alpha^{(1,0)} | \Psi_k \rangle \langle \Psi_k | \mathcal{H}_{A\beta}^{(0,1)} | \Psi_0 \rangle + \langle \Psi_0 | \mathcal{H}_{A\alpha}^{(0,1)} | \Psi_k \rangle \langle \Psi_k | \mathcal{H}_\beta^{(1,0)} | \Psi_0 \rangle]. \tag{3.8}$$

Equation (3.8) may be evaluated by introducing explicit expressions for the operators:

$$\mathcal{H}_{A\alpha\beta}^{(1,1)} = \frac{\mu_0 e^2}{8\pi m} \sum_i (\mathbf{r}_i \cdot \mathbf{r}_{iA} \delta_{\alpha\beta} - r_{i\alpha} r_{iA\beta}) r_{iA}^{-3}, \tag{3.9}$$

$$\mathcal{H}_\alpha^{(1,0)} = \frac{\mu_0 e \hbar}{8\pi m} \sum_i L_{i\alpha}, \tag{3.10}$$

$$\mathcal{H}_{A\alpha}^{(0,1)} = \frac{\mu_0 e}{4\pi m} \sum_i L_{iA\alpha} r_{iA}^{-3}, \tag{3.11}$$

where μ_0 is the permittivity of free space, r_{iA} refers to the separation of electron i from nucleus A, $\delta_{\alpha\beta}$ is the Kronecker delta and L_i is the electron orbital angular momentum operator.

The resulting shielding expression may be written as

$$\sigma_{A\alpha\beta} = \sigma_{A\alpha\beta}^{d} + \sigma_{A\alpha\beta}^{p}, \tag{3.12}$$

where the diamagnetic and paramagnetic contributions respectively are given by

$$\sigma_{A\alpha\beta}^{d} = \frac{\mu_0 e^2}{8\pi m} \langle \Psi_0 | \sum_i (\mathbf{r}_i \cdot \mathbf{r}_{iA} \delta_{\alpha\beta} - r_{i\alpha} r_{iA\beta}) r_{iA}^{-3} | \Psi_0 \rangle \tag{3.13}$$

and

$$\sigma_{A\alpha\beta}^{p} = -\frac{\mu_0 e^2}{8\pi m^2} \sum_{k \neq 0} (E_k - E_0)^{-1} [\langle \Psi_0 | \sum_i L_{i\alpha} | \Psi_k \rangle$$
$$\times \langle \Psi_k | \sum_i L_{iA\beta} r_{iA}^{-3} | \Psi_0 \rangle$$
$$+ \langle \Psi_0 | \sum_i L_{iA\alpha} r_{iA}^{-3} | \Psi_k \rangle \langle \Psi_k | \sum_i L_{i\beta} | \Psi_0 \rangle]. \tag{3.14}$$

Equations (3.12)–(3.14) constitute Ramsey's expression for nuclear shielding. It is noteworthy that the diamagnetic and paramagnetic contributions act in opposite directions as indicated by the negative sign of the expression for $\sigma^P_{A\alpha\beta}$. Since the paramagnetic term requires the presence of electrons with orbital angular momentum, it will not be operative for s valence electrons. Thus only the diamagnetic term contributes to the ground state shielding of hydrogen nuclei, which accounts for the small chemical shift range exhibited by hydrogen and other nuclei with a similar outer electronic configuration.

The names diamagnetic and paramagnetic used in connection with the expressions (3.13) and (3.14), respectively, arise from their similarity to those in the Van Vleck formulation of the magnetic susceptibility tensor.[4] Care is required in using these terms in discussions of nuclear shielding. Thus the shielding of nuclei in molecules which are formally diamagnetic is describable by means of diamagnetic and paramagnetic contributions. Whilst for paramagnetic species the nuclear shielding depends upon the unpaired electrons which give rise to contact and so-called pseudo-contact interactions.[5]

Returning to Ramsey's expression it is clear that the diamagnetic term arises from the ground state electronic function. As such it is the molecular counterpart of the Lamb formula for atomic shielding. In contrast to this the paramagnetic term depends upon both the ground and excited electronic states which exist in the absence of an applied magnetic field. A major difficulty involved in the evaluation of the paramagnetic term is that, in general, little is known of the functions describing the high energy discrete states. In addition, the summation in equation (3.14) includes integrations over continuum states about which even less is known than for the discrete states. It seems likely that the shielding contributions from the continuum states may be more important than some of those from discrete states.[6]

A more general problem which arises from Ramsey's model is that the distances, r, mentioned in equations (3.13) and (3.14) are referred to an arbitrary origin, thus the value of $\sigma_{A\alpha\beta}$ calculated depends upon the origin chosen for the calculation in question unless the basis set used is a complete one. However, the diamagnetic and paramagnetic contributions to the shielding tensor remain gauge dependent even when a complete basis set is employed.

Gauge-dependent contributions arise when an incomplete basis set is used due to the supposition of a common origin, which is a poor one for an orbital whose centroid is far removed from that origin. Thus a complete basis set is required to correct for the spurious shielding contributions which are produced whenever the gauge origin is inappropriately chosen.

This problem of gauge dependence raises a number of difficulties.[6-8] Amongst these difficulties is the fundamental physical requirement that nuclear shielding be independent of gauge, because it manifestly does not depend upon the means of accurate measurement employed in an experiment. A means of effectively removing the gauge problem from consideration in nuclear shielding calculations is discussed in Section 3.C.

Further difficulties with Ramsey's model arise from the fact that the diamagnetic and paramagnetic terms tend to increase in an antiparallel fashion as the molecular size increases. Thus the total shielding is expressed by the difference between two fairly large terms. Such a situation is unsatisfactory in that it can give rise to considerable errors. Objections of this kind can be removed by the development of more localized shielding expressions, as discussed in Sections 3.B and 3.C.

Some simplification of equation (3.14) is available by choosing an average excitation energy, ΔE, for the excited states and invoking the closure approximation. When applied to equation (3.14) this gives

$$\sigma^p_{A\alpha\beta} = -\frac{\mu_0 e^2}{4\pi m^2 \, \Delta E} \left[\langle \Psi_0 | \sum_i L_{i\alpha} \sum_i L_{iA\beta} r_i^{-3} | \Psi_0 \rangle \right]. \qquad (3.15)$$

This simplification of the paramagnetic contribution to the shielding has the advantage of only requiring a knowledge of the ground state eigenfunctions. Estimates of these may be made at various levels of MO approximation, as discussed in Chapter 2. However, there is no satisfactory means of finding a suitable value for ΔE by independent procedures. Consequently, ΔE is usually considered to be an additional empirical parameter; thus, from the quantum mechanical point of view, this approach leaves something to be desired.

It is perhaps instructive at this juncture to appreciate that the requirement of some knowledge of the excited states, in dealing with nuclear shielding, arises as an artefact of the second-order perturbation term given by equation (3.7). The expression for the excited states arises from the approximations implicit in the Rayleigh–Schrödinger perturbation theory, Consequently, if it were possible to evaluate the many-body problem exactly, there would be no recourse to a consideration of excited electronic states. Thus there is no strong reason to anticipate any relationship between ΔE and observed electronic transition energies for the molecule in question.

The difficulties associated with a satisfactory estimation of the excitation energies may be resolved by an alternative, but equivalent, quantum mechanical procedure for formulating the shielding tensor elements. This involves the use of finite perturbation theory (FPT)[9,10] rather than the Rayleigh–Schrödinger sum over states (SOS) theory used in the derivation of equation (3.5).

Within the FPT approach,[6] components of the shielding tensor are evaluated from

$$\sigma_{A\alpha\beta} = \left[\frac{\partial^2 E(\mu_A, B)}{\partial \mu_{A\beta} \partial B_\alpha}\right]_{\mu_{A\beta} = B_\alpha = 0} . \quad (3.16)$$

The FPT method considers a perturbed Hamiltonian, $\mathcal{H}(\lambda)$, expressed in terms of a number of external perturbations $\lambda_1, \lambda_2, \ldots$ and the corresponding operators $\mathcal{H}_1, \mathcal{H}_2, \ldots$. Thus

$$\mathcal{H}(\lambda) = \mathcal{H}^0 + \lambda_1 \mathcal{H}_1 + \lambda_2 \mathcal{H}_2 + \ldots . \quad (3.17)$$

From the Hellmann–Feynmann theorem it can be shown that, as in Section 2.A.3 and equation (2.40),

$$\frac{\partial E(\lambda)}{\partial \lambda_1} = \langle \Psi(\lambda)| \frac{\partial \mathcal{H}(\lambda)}{\partial \lambda_1} |\Psi(\lambda)\rangle \quad (3.18)$$

$$= \langle \Psi(\lambda)|\mathcal{H}_1|\Psi(\lambda)\rangle; \quad (3.19)$$

consequently,

$$\left[\frac{\partial^2 E(\lambda_1, \lambda_2)}{\partial \lambda_1 \partial \lambda_2}\right]_{\lambda_1 = \lambda_2 = 0} = \left[\frac{\partial}{\partial \lambda_2} \langle \Psi(\lambda_2)|\mathcal{H}_1|\Psi(\lambda_2)\rangle\right]_{\lambda_2 = 0} \quad (3.20)$$

Hence, to calculate $\sigma_{A\alpha\beta}$ by the FPT procedure, $\Psi(B)$ is calculated from a Hamiltonian of the form

$$\mathcal{H}(B) = \mathcal{H}^0 + B\mathcal{H}_2. \quad (3.21)$$

$\sigma_{A\alpha\beta}$ is obtained as the first derivative of the expectation value of $\mathcal{H}_1$ over $\Psi(B)$ evaluated numerically at $B = 0$. Thus,

$$\sigma_{A\alpha\beta} = \left[\frac{\partial}{\partial B_\alpha} \langle \Psi(B_\alpha)|\mathcal{H}_{A\beta}^{(0,1)} + \mathcal{H}_{A\alpha\beta}^{(1,1)}|\Psi_1(B_\alpha)\rangle\right]_{B_\alpha = 0} . \quad (3.22)$$

Equation (3.22) may be rewritten in terms of first- and second-order contributions as

$$\sigma_{A\alpha\beta} = \langle \Psi_0|\mathcal{H}_{A\alpha\beta}^{(1,1)}|\Psi_0\rangle$$

$$+ \left[\frac{\partial}{\partial B_\alpha} \langle \Psi(B_\alpha)|\mathcal{H}_{A\beta}^{(0,1)}|\Psi(B_\alpha)\rangle\right]_{B_\alpha = 0}, \quad (3.23)$$

where expressions for $\mathcal{H}_{A\alpha\beta}^{(1,1)}$ and $\mathcal{H}_{A\beta}^{(0,1)}$ are provided by equations (3.9) and (3.11) respectively.

The first term on the right-hand side of equation (3.23) is the same as in equation (3.8) and represents the diamagnetic contribution to the nuclear shielding. The paramagnetic term in equation (3.23) is determined solely by the perturbed function, $\Psi(B_\alpha)$, and thus differs significantly from its counterpart in equation (3.8). The absence of excitation energies in equation (3.23) is noteworthy.

So far it has been tacitly assumed that the Schrödinger equation is solvable for a many electron, polyatomic molecule. As discussed in Chapter 2, this is only possible in approximate form for practical purposes.

3.B CALCULATIONS BASED UPON SCF THEORIES

The LCAO-SCF equations are developed in Section 2.B.3. When applied to equations (3.13) and (3.14) the SOS expressions for the diamagnetic and paramagnetic contributions to the shielding tensor components are given by

$$\sigma^d_{A\alpha\beta} = \frac{\mu_0 e^2}{8\pi m} \sum_j^{occ.} \langle \varphi_j | (\mathbf{r} \cdot \mathbf{r}_A \delta_{\alpha\beta} - r_\alpha r_{A\beta}) r_A^{-3} | \varphi_j \rangle \qquad (3.24)$$

and

$$\sigma^p_{A\alpha\beta} = -\frac{\mu_0 e^2}{8\pi m^2} \sum_j^{occ.} \sum_k^{unocc.} (E_k - E_j)^{-1}$$

$$\times [\langle \varphi_j | L_\alpha | \varphi_k \rangle \langle \varphi_k | L_{A\beta} r_A^{-3} | \varphi_j \rangle + \langle \varphi_j | L_{A\alpha} r_A^{-3} | \varphi_k \rangle \langle \varphi_k | L_\beta | \varphi_j \rangle], \quad (3.25)$$

where φ_j and φ_k refer to occupied and unoccupied MOs respectively, with corresponding energies E_j and E_k. Thus $\sigma^d_{A\alpha\beta}$ and $\sigma^p_{A\alpha\beta}$ are reduced to integrals over MOs which pertain to the unperturbed state of the molecule under consideration. In contrast to equation (3.14), equation (3.25) for the paramagnetic term involves a summation over a finite number of MOs only. However, the description of excited states by unoccupied MOs is usually a poor one such that contributions to $\sigma^p_{A\alpha\beta}$ from higher excited states may be in considerable error. This is particularly true if the basis set is restricted to the valence atomic shell.

The paramagnetic shielding contribution may be considered to arise from a charge circulation effect. This is produced by the nature of the orbital angular momentum operators, L, in the matrix elements present in equation (3.25). For example in the case of p atomic orbitals, $L_z|p_x\rangle = -|p_y\rangle$, $L_z|p_y\rangle = |p_x\rangle$ and $L_z|p_z\rangle = 0$, with similar cyclic relationships for the operators L_x and L_y.[7] Thus the non-zero contributions to the shielding are produced by those orbital angular momentum matrix elements which account for the rotation of a filled p orbital into the space previously occupied by an empty p orbital, i.e. charge rotation.

Equations (3.24) and (3.25) may be used with some success in dealing with diatomic molecules. The remaining gauge problem can be counteracted by arbitrarily taking the gauge origin to be coincident with the nucleus of interest.

In the corresponding FPT approach, equation (3.23) requires $\Psi(B_\alpha)$ and the derivative of the expectation value of $\mathscr{H}_{A\beta}^{(0,1)}$ for the calculation of the paramagnetic term. From equation (3.21) it follows that $\mathscr{H}(B_\alpha)$ for use in the perturbed eigenfunction calculation is given by

$$\mathscr{H}(B_\alpha) = \mathscr{H}^0 + B_\alpha \mathscr{H}_\alpha^{(1,0)}. \tag{3.26}$$

Thus magnetic shielding tensors for all nuclei in a given molecule can be found from three perturbed eigenfunction calculations, corresponding to $\alpha = x, y$ and z. A SCF procedure for determining $\Psi(B)$ in a MO description can be developed by means of a variational procedure. The LCAO coefficients, $C_{\mu j}$, given in equation (2.54) are allowed to be complex due to the pure imaginary nature of the perturbation. This is achieved by introducing expansions of $C_{\mu j}$ and the bond-order matrix elements $P_{\mu\nu}$ (defined by equation (2.59)), similar to that of equation (3.4):

$$C_{\mu j}(B_\alpha, \mu_{A\alpha}) = C_{\mu j}^{(0)} + iB_\alpha (C_j^{(1,0)})_\alpha + i\mu_{A\alpha}(C_{A,\mu j}^{(0,1)})_\alpha + \ldots \tag{3.27}$$

and

$$P_{\mu\nu}(B_\alpha, \mu_{A\mu}) = P_{\mu\nu}^{(0)} + iB_\alpha (P_{\mu\nu}^{(1,0)})_\alpha + i\mu_{A\alpha}(P_{A\mu\nu}^{(0,1)})_\alpha + \ldots. \tag{3.28}$$

In equations (3.27) and (3.28) i stands for $\sqrt{-1}$ and

$$P_{\mu\nu}^{(0)} = 2 \sum_j^{\text{occ.}} C_{\mu j}^{(0)} C_{\nu j}^{(0)}. \tag{3.29}$$

Similarly,

$$(P_{\mu\nu}^{(1,0)})_\alpha = 2 \sum_j^{\text{occ.}} [C_{\mu j}^{(0)}(C_{\nu j}^{(1,0)})_\alpha - (C_{\mu j}^{(1,0)})_\alpha C_{\nu j}^{(0)}]. \tag{3.30}$$

The resulting MO description of equation (3.23) for the FPT account of the nuclear shielding components is given by

$$\sigma_{A\alpha\beta} = \frac{\mu_0 e^2}{8\pi m} \sum_\mu^A \sum_\nu^A P_{\mu\nu}^{(0)} \langle \chi_\mu | (\mathbf{r} \cdot \mathbf{r}_A \delta_{\alpha\beta} - r_\alpha r_{A\beta}) r_A^{-3} | \chi_\nu \rangle$$

$$- \frac{\mu_0 e}{8\pi m^2} \sum_\mu^A \sum_\nu^A (P_{\mu\nu}^{(1,0)})_\alpha \langle \chi_\mu | L_{A\beta} r_A^{-3} | \chi_\nu \rangle. \tag{3.31}$$

Equation (3.31) may be evaluated at any LCAO-SCF level of approximation. This type of approach was developed by Stevens and Lipscomb for diatomic molecules.[11-14] An equivalent approach was later proposed by Ditchfield *et al.*[15,16]

Due to the necessity of diagonalizing a complex matrix in the FPT procedure the calculations of nuclear shielding can be rather time consuming.

Hence *ab initio* calculations for large molecules by the FPT approach could be prohibitively expensive. A somewhat cheaper alternative, at this level of approximation, could be the SOS method including configuration interaction (CI). The presence of CI in a calculation is known to provide a more realistic account of excitation energies.

Calculations involving the 4-31G basis set within the SOS plus CI and FPT frameworks for the heavy atoms in HF, H_2O, NH_3 and CH_4 produce shielding results which differ by less than 10%.[17] In those cases where CI is important for the unperturbed functions used in the SOS calculations, the results from the SOS, without CI, and the FPT calculations differ significantly. The results obtained point to the equivalence of the two perturbation procedures for calculating nuclear shielding and could indicate the route of future computational developments in this area.

Shielding components, $\sigma_{A\alpha\beta}$, obtained from equation (3.31) are in principle only independent of gauge if a complete set of atomic functions χ, is used in the calculation. For small molecules it may be practicable to employ an extended basis set in the calculation. Under such circumstances the calculated shieldings for nuclei in similar positions relative to the origin of the magnetic field vector are in reasonable agreement with experiment. However, for nuclei in markedly different positions relative to the origin the calculated chemical shifts are less satisfactory.

From the foregoing account it seems that the gauge problems attached to the SCF-MO description of nuclear shielding render the approach rather unattractive for a general theory of chemical shifts. In order to deal effectively with molecules of a reasonable size it would be instructive to have a theory which compounds the shielding from a number of localized contributions. Such an approach would have the merit of permitting the nuclear shielding to be discussed in terms of chemically interesting aspects such as atomic and bonding contributions. That such would be a reasonable development in the theory of nuclear shielding is reinforced by a consideration of the relationship between the diamagnetic and paramagnetic contributions to the nuclear shielding and molecular magnetic susceptibility, χ, tensors. Namely that

$$\sigma^d \propto \chi^d/Nr_k^3 \qquad (3.32)$$

and

$$\sigma^p \propto \chi^p/Nr_k^3, \qquad (3.33)$$

where N is the Avogadro number and r_k refers to the separation between the nucleus of interest and the kth electron. It is clear from equations (3.32) and (3.33) that the nuclear shielding expressions are dominated by contributions from electrons which are relatively close to the nucleus.

Before considering this aspect, and the question of gauge dependence, further, it is of interest to note that the paramagnetic part of the shielding tensor is closely related to the molecular spin–rotation tensor,

$$\sigma_A^p = \frac{\mu_0 e^2 C \cdot I}{8\pi m \mu_N \gamma_A M} - \frac{\mu_0 e^2}{12\pi m} \sum_B{}' \frac{Z_B}{r_B}, \qquad (3.34)$$

where M denotes the proton mass, μ_N is the nuclear magneton and Z_B is the atomic number of nucleus B which is at a distance r_B from nucleus A. The prime on the summation over B indicates that the shielded nucleus in question is omitted, C and I are the rotationally averaged values of the spin–rotation tensor and the molecular moment of inertia respectively.

Thus values of C taken from molecular beam measurements can be used to obtain an estimate of the paramagnetic contribution to the nuclear shielding.[18] A similar expression[19] for σ_A^d is

$$\sigma_A^d \simeq \sigma_{(\text{free atom})}^d + \frac{\mu_0 e^2}{12\pi m} \sum_B{}' \frac{Z_B}{r_B}. \qquad (3.35)$$

Values of σ^d (free atom) are readily available[20] and data for some commonly encountered NMR nuclei are given in Table 3.1. Consequently equations (3.34) and (3.35) together provide an empirical estimate of the nuclear shielding tensor. This procedure is employed to give the experimental results for the ^{31}P nucleus of PH_3 in Table 3.2. Although, in general, the rotationally averaged shielding values obtained by this procedure are satisfactory, the estimates of shielding anisotropy are less reliable due to approximations introduced in the derivation of equations (3.34) and (3.35).

Table 3.1 Some free atom values of σ^d

Nucleus	σ^d(p.p.m.)
B	201.99
C	260.74
N	325.47
O	395.11
F	470.71
Na	628.90
Al	789.88
Si	874.09
P	961.14
Cl	1 142.64
K	1 329.36
Br	3 121.19
I	5 501.64
Hg	9 729.07
Tl	9 894.16
Pb	10 060.92

Table 3.2 ^{31}P shielding data for PH$_3$: calculations employ the equilibrium configuration with the gauge origin on phosphorus

σ_{av}^{d}(p.p.m.)	σ_{av}^{p}(p.p.m.)	σ_{av}^{total}(p.p.m.)	Source of data and references
981.35	−403.79	577.56	Ab initio FPT including only symmetrically distinct two-electron integrals:[21] 112 primitives and 91 contracted functions used
980.89	−391.27	589.62	Ab initio FPT, well polarized basis set: 92 primitives and 83 contracted functions[22]
980.99	−347.48	633.51	Pseudo-potential for inner shells: 57 primitives and 33 contracted functions[23]
984.0	−387.0	597.0	Experimental spin rotation results[17]

The presence of the same final term, with opposite signs, in equations (3.34) and (3.35) ensures that the diamagnetic and paramagnetic terms are dependent upon molecular size, as noted in calculations based on Ramsey's formulation of the shielding tensor. This final term also mimics the gauge dependence of the shielding contributions, which may still arise even when the employment of a complete set of atomic orbitals entails that the calculated value of the total shielding is gauge independent.

As shown in Table 3.2 the calculated value of σ^d for the ^{31}P nucleus in PH$_3$ is in reasonable agreement with the experimental value. However, the value of σ^p depends closely on the quality of the basis set used in the *ab initio* calculations. As the quality of the basis set improves, the paramagnetic term becomes more accurate and the gauge dependence of the shielding becomes less pronounced.

Unfortunately, it is not practicable to use very large basis sets for other than small molecules; thus the gauge problem remains. Chemical interests in nuclear shielding reside in chemical shifts either for different nuclei in the same molecule or for given nuclei in a series of molecules, rather than in absolute shielding measurements. Results which are gauge dependent can produce difficulties when series of chemical shift values are under consideration.

3.C CALCULATIONS GIVING GAUGE-INDEPENDENT SHIELDING DATA

The SCF shielding formulations discussed above are only capable of providing gauge-independent results when complete sets of atomic functions,

χ, are used in the calculation. For molecules of a reasonable size an unjustifiable amount of computing time and power would be necessary for calculations employing complete basis sets.

A move towards obtaining gauge-independent shielding data is thus best made in a different direction. The remedy most frequently encountered is to use a linear combination of gauge-invariant atomic orbitals (GIAOs), θ_μ, to describe each MO. It is unfortunate that these orbitals are so named, since from equation (3.36) they are clearly gauge dependent. The accepted GIAO nomenclature is used here with the understanding that it is not strictly correct. The shielding values obtained by the use of GIAOs are gauge independent:

$$\theta_\mu(B) = \chi_\mu \exp\left[-\left(\frac{ie}{\hbar}\right)\mathbf{A}_\mu \cdot \mathbf{r}\right], \quad (3.36)$$

where

$$\mathbf{A}_\mu = \tfrac{1}{2}\mathbf{B} \times \mathbf{r}_\mu. \quad (3.37)$$

Thus the real atomic orbitals, χ_μ, are multiplied by a factor which is dependent upon the gauge of the vector potential $\mathbf{A}$, whose value at the nuclear position $\mathbf{r}_\mu$ of the atom containing χ_μ is $\mathbf{A}_\mu$, and $\mathbf{r}$ is the position vector for an electron in χ_μ. The use of the orbitals θ_μ ensures that the resulting expressions for the diamagnetic and paramagnetic shielding contributions are localized ones due to the presence in the exponential function of $(-\mathbf{r})$. In addition, the inclusion of gauge in the exponent results in the removal of the gauge when the orbitals θ_μ are used to evaluate matrix elements, including the operators $\mathcal{H}^{(1,1)}_{A\alpha\beta}$ and $\mathcal{H}^{(0,1)}_{A\alpha}$ from equations (3.9) and (3.11) respectively.

By comparison with equations (3.3) and (3.4), the total energy of the system may be expressed as a perturbation expansion in $\mathbf{A}$, where the total field at electron j is considered as a combination of $\mathbf{B}$ and the field of $\boldsymbol{\mu}_A$.

$$\mathbf{A} = \tfrac{1}{2}\mathbf{B} \times \mathbf{r}_j + (\mu_0/4\pi)(\boldsymbol{\mu}_A \times \mathbf{r}_{jA})r_{jA}^{-3}. \quad (3.38)$$

The shielding is obtained from the coefficient of the term in the energy expression, cf. equation (3.3), which is bilinear in $\mathbf{B}$ and $\boldsymbol{\mu}_A$.

3.C.1 FPT Calculations

By means of GIAOs Ditchfield[24] has derived an expression for the elements of the shielding tensor within the FPT framework:

$$\sigma_{A\alpha\beta} = \frac{\mu_0 e^2}{8\pi m} \sum_\mu^A \sum_\nu^A P^{(0)}_{\mu\nu} \langle \chi_\mu | (\mathbf{r}_A \cdot \mathbf{r}_\nu \delta_{\alpha\beta} - \mathbf{r}_{A\alpha} \cdot \mathbf{r}_{\nu\beta})r_A^{-3} | \chi_\nu \rangle$$

$$+ \frac{\mu_0 e \hbar^2}{8\pi m} \sum_\mu^A \sum_\nu^A \left\{ P_{\mu\nu}^{(0)} \left[\left\langle \frac{\partial \theta_\mu}{\partial B_\alpha} \left| \frac{L_{A\beta}}{r_A^3} \right| \chi_\nu \right\rangle \right. \right.$$

$$\left. \left. + \left\langle \chi_\mu \left| \frac{L_{A\beta}}{r_A^3} \right| \frac{\partial \theta_\nu}{\partial B_\alpha} \right\rangle \right] + \frac{\partial P_{\mu\nu}(B_\alpha)}{\partial B_\alpha} \left\langle \chi_\mu \left| \frac{L_{A\beta}}{r_A^3} \right| \chi_\nu \right\rangle \right\}, \quad (3.39)$$

where $P_{\mu\nu}^{(0)}$ is defined by equation (3.29). By comparison with equation (3.31) it is apparent that the first term of equation (3.39) gives the diamagnetic contribution to the shielding tensor while the remaining terms describe the paramagnetic component. By means of equation (3.39) the shielding tensors for all nuclei in a given molecule may be obtained, at any MO level of approximation, from three perturbed eigenfunction calculations corresponding to $\alpha = x, y$ or z.

Minimum basis set calculations at the *ab initio* level, using a least squares fitted sum of five Gaussians (STO-5G) to replace each STO, provide shielding results from equation (3.39) which are both gauge independent and in good agreement with information obtained from spin–rotation experiments.[24]

Although equation (3.39) provides a satisfactory estimate of nuclear shielding it gives little indication of the mechanisms by which shielding occurs. Ditchfield[6,25] has shown that such information may be obtained by decomposing the shielding expression into a number of local, non-local and interatomic contributions, such that

$$\sigma_A = \sigma_A^d(\text{loc.}) + \sigma_A^p(\text{loc.}) + \sigma_A^d(\text{non-loc.}) + \sigma_A^p(\text{non-loc.})$$
$$+ \sigma_A^d(\text{inter.}) + \sigma_A^p(\text{inter.}). \quad (3.40)$$

The local diamagnetic and paramagnetic terms in equation (3.40) arise from shielding contributions due to electrons on atom A, the non-local terms are produced by electrons on neighbouring nuclei and the interatomic contributions arise from delocalized electrons such as those which produce ring currents.

The various diamagnetic and paramagnetic terms comprising equation (3.40) are not directly comparable to those bearing the same names in Ramsey's theory. Care must be exercised to avoid the confusion which could arise from any comparison of this kind.

For molecules containing two first row atoms equation (3.39) provides reasonable shielding results with the minimal STO-5G and slightly extended basis sets in a modest amount of computing time. More satisfactory results are obtained with the extended 4-31G basis set which requires a larger amount of computational effort. For larger molecules the necessity of computing a large number of two-electron integrals between GIAOs renders equation (3.39) impracticable.

In a study of the ^{31}P shielding of phosphate groups in a dinucleoside monophosphate and in a phospholipid, the calculated shieldings were obtained for dimethyl phosphate, monomethyl phosphate and the orthophosphoric acid monoanion as model compounds for the biological phosphate group.[26] The calculations use a minimal basis set which requires the evaluation of 142 two-electron integrals for dimethyl phosphate rather than the 180 necessary when the 4-31G basis set is used. Even with this approximation to reality an extensive amount of computer time is required.

Considerable simplification, and reduction in demands on computer power, are forthcoming when semi-empirical MO methods are employed. However, the use of these methods inevitably leads to a dissipation of some of the rigour involved in the derivation of equation (3.39). Against this it should be remembered that CNDO/INDO calculations are well established and can readily be applied to molecules of a reasonable size.[27]

Application of semi-empirical MO approximations to the local and non-local terms expressed in equation (3.40) leads to[28]

$$\sigma_A = \sigma_A^{dAA} + \sigma_A^{pAA} + \sum_{C \neq A} \left[\sigma_A^{dAC} + \sigma_A^{pAC} \right. \\ \left. + \sigma_A^{dCC} + \sigma_A^{pCC} + \sum_{D \neq A \neq C} (\sigma_A^{dDC} + \sigma_A^{pDC}) \right]. \quad (3.41)$$

The first two terms in equation (3.41) refer to the one-centre diamagnetic and paramagnetic shielding contributions respectively, the following four terms arise from shielding currents on atoms A and C, thus these are two-centre terms. The final two contributions to equation (3.41) are due to the effect of the induced current density, between atoms D and C, on the shielding of nucleus A; consequently these are three-centre terms. Explicit expressions for these shielding contributions are as follows:

$$\sigma_{A\alpha\beta}^{dAA} = \frac{\mu_0 e^2}{8\pi m} \sum_\mu^A \sum_\nu^A P_{\mu\nu} \langle \chi_\mu | (\mathbf{r}_A \cdot \mathbf{r}_A \delta_{\alpha\beta} - r_{A\alpha} \cdot r_{A\beta}) r_A^{-3} | \chi_\nu \rangle, \quad (3.42)$$

$$\sigma_{A\alpha\beta}^{pAA} = \frac{\mu_0 e^2}{4\pi m^2} \sum_\mu^A \sum_\nu^A \frac{\partial P_{\mu\nu}(B_\beta)}{\partial B_\beta} \langle \chi_\mu | \frac{L_{A\alpha}}{r_A^3} | \chi_\nu \rangle, \quad (3.43)$$

$$\sigma_{A\alpha\beta}^{dCC} = \frac{\mu_0 e^2}{8\pi m} \sum_\mu^C \sum_\nu^C P_{\mu\nu} \langle \chi_\mu | (\mathbf{r}_C \cdot \mathbf{r}_A \delta_{\alpha\beta} - r_{C\alpha} r_{A\beta}) r_A^{-3} | \chi_\nu \rangle, \quad (3.44)$$

$$\sigma_{A\alpha\beta}^{pCC} = \frac{\mu_0 e^2}{4\pi m^2} \sum_\mu^C \sum_\nu^C \frac{\partial P_{\mu\nu}(B_\beta)}{\partial B_\beta} \langle \chi_\mu | \frac{L_{A\alpha}}{r_A^3} | \chi_\nu \rangle, \quad (3.45)$$

$$\sigma_{A\alpha\beta}^{dAC} = \frac{\mu_0 e^2}{8\pi m} \sum_\mu^A \sum_\nu^C P_{\mu\nu} \langle \chi_\mu | (\mathbf{r}_C \cdot \mathbf{r}_A \delta_{\alpha\beta} - r_{C\alpha} r_{A\beta}) r_A^{-3} | \chi_\nu \rangle, \quad (3.46)$$

$$\sigma_{A\alpha\beta}^{pAC} = \frac{\mu_0 e^2}{4\pi m^2} \sum_\mu^A \sum_\nu^C \frac{\partial P_{\mu\nu}(B_\beta)}{\partial B_\beta} \langle \chi_\mu | \frac{L_{A\alpha}}{r_A^3} | \chi_\nu \rangle$$

$$- \sum_\mu^A \sum_\nu^C (\mathbf{R}_C \times \mathbf{R}_A)_\beta P_{\mu\nu} \langle \chi_\mu | \frac{L_{A\alpha}}{r_A^3} | \chi_\nu \rangle$$

$$+ \sum_\mu^A \sum_\nu^C P_{\mu\nu} \langle (\mathbf{R}_{AC} \times \mathbf{r}_A)_\beta \chi_\mu | \frac{L_{A\alpha}}{r_A^3} | \chi_\nu \rangle, \qquad (3.47)$$

$$\sigma_{A\alpha\beta}^{dDC} = \frac{\mu_0 e^2}{8\pi m} \sum_\mu^D \sum_\nu^C P_{\mu\nu} \langle \chi_\mu | (\mathbf{r}_C \cdot \mathbf{r}_A \delta_{\alpha\beta} - r_{C\alpha} r_{A\beta}) r_A^{-3} | \chi_\nu \rangle, \qquad (3.48)$$

$$\sigma_{A\alpha\beta}^{pDC} = \frac{\mu_0 e^2}{4\pi m^2} \sum_\mu^D \sum_\nu^C \frac{\partial P_{\mu\nu}(B_\beta)}{\partial B_\beta} \langle \chi_\mu | \frac{L_{A\alpha}}{r_A^3} | \chi_\nu \rangle$$

$$- \sum_\mu^D \sum_\nu^C (\mathbf{R}_C \times \mathbf{R}_D)_\beta P_{\mu\nu} \langle \chi_\mu | \frac{L_{A\alpha}}{r_A^3} | \chi_\nu \rangle$$

$$+ \sum_\mu^D \sum_\nu^C P_{\mu\nu} \langle (\mathbf{R}_{DC} \times \mathbf{r}_D)_\beta \chi_\mu | \frac{L_{A\alpha}}{r_A^3} | \chi_\nu \rangle. \qquad (3.49)$$

Equations (3.47) and (3.49) contain an explicit origin dependence through the terms in ($\mathbf{R}_C \times \mathbf{R}_A$) and ($\mathbf{R}_C \times \mathbf{R}_D$). These are added to remove the origin dependence within the expression for $\partial P_{\mu\nu}(B_\beta)/\partial B_\beta$, where $\mathbf{R}_A$ refers to the position vector for nucleus A with respect to the origin of the magnetic field and $\mathbf{r}_A$ relates to the separation of the electrons from nucleus A. Thus the shielding expressions represented by equations (3.42) to (3.49) are all gauge independent.

INDO parameterized calculations of nuclear shielding using a truncated two-centre Hamiltonian have been reported.[29] The approximations involved are tantamount to including only $\sigma_{A\alpha\beta}^{dAA}$, $\sigma_{A\alpha\beta}^{pAA}$, $\sigma_{A\alpha\beta}^{dCC}$ and $\sigma_{A\alpha\beta}^{pCC}$ of the various contributions described by equations (3.42)–(3.49). The values of $\sigma_{A\alpha\beta}^{dCC}$ and $\sigma_{A\alpha\beta}^{pCC}$ were estimated by assuming that the effects of the currents on atom C, as far as the shielding of nucleus A is concerned, can be approximated as arising from a dipole. These calculations are unsatisfactory in that they require non-standard values of the INDO parameters used to estimate the matrix elements given by equations (2.72) and (2.73), as well as some optimization of the Slater exponents.[27,28]

A measure of the reliability of any MO theory is its ability to predict a number of different molecular properties. Thus it is somewhat disconcerting that non-standard parameter sets are required to obtain a reasonable account of nuclear shielding.

Intuitively the three-centre terms, equations (3.48) and (3.49), are most

Table 3.3 Calculated and experimental chemical shifts (in parts per million) for some representative boron-containing compounds

Compound	σ_A^{AA}	σ_A^{CC}	σ_A^C	σ_A^{CD}	σ_A	δ_A(calc.)	δ_A(exp.)
B_5H_{11} (B_1)	30.55	4.86	25.81	29.17	90.39	−81.03	−72.8
*B_5H_9 (B_1)	17.70	4.44	17.66	43.68	74.49	−65.13	−70.2
*B_4H_{10} (B_1)	9.10	3.74	26.63	22.57	62.04	−52.68	−59.3
$B_{10}H_{14}$ (B_2)	19.54	6.19	16.85	35.31	77.89	−68.53	−53.3
*B_5H_9 (B_2)	−2.70	4.22	23.85	23.36	48.45	−39.09	−30.6
*B_4H_{10} (B_2)	−4.42	3.38	28.18	14.16	42.31	−32.95	−24.4
B_5H_{11} (B_3)	−10.32	3.43	26.68	13.13	32.92	−23.56	−17.0
$B_{10}H_{14}$ (B_5)	−4.96	4.73	18.05	25.36	43.19	−33.83	−16.8
B_5H_{11} (B_2)	−11.61	4.73	20.33	20.45	33.90	−24.54	−10.0
$B_{10}H_{14}$ (B_6)	−15.57	4.24	19.15	21.52	29.34	−19.98	−7.8
$B_{10}H_{14}$ (B_6)	−10.23	5.99	9.19	33.15	38.10	−28.74	−6.2
*B_2H_6	−25.40	2.82	26.43	5.51	9.36	0.00	0.0
$B(C_2H_3)_3$	−36.66	−2.96	5.24	1.97	−32.40	41.76	37.7
$(CH_3)_2BC_3H_3$	−57.84	1.99	0.69	6.06	−49.10	58.46	53.5
$B(CH_3)_3$	−69.25	−0.46	−1.44	10.88	−60.26	69.62	68.5

likely to contribute to the shielding in a significant manner if atoms A, C and D are linearly bonded with some multiple bonding between C and D. Thus the methyl carbon of methylacetylene is more likely to have a three-centre shielding contribution than is a methyl carbon in propane.

The significance of the various terms in equation (3.41) has been ascertained at the INDO level by some calculations on electron-deficient boron hydrides[30] (Table 3.3). In this investigation all of the shielding contributions given by equations (3.42)–(3.49) are calculated explicitly. The dipole approximation is not used for the multi-centre terms.

It is clear from the results given in Table 3.3 that the one-, two- and three-centre shielding terms all make significant contributions. Thus current densities remote from the nucleus of interest have a strong influence in determining the boron shieldings in the molecules considered.

This result appears to open the way to fairly reliable chemical shift calculations, at the semi-empirical level, on molecules which are sufficiently large to be of widespread chemical interest. However, the point must be made that optimized values are used for the parameters involved in estimating the off-diagonal matrix elements given by equation (2.73). The molecules marked with an asterisk in Table 3.3 are involved in the optimization procedure. Variation of the INDO parameters occurs in order to obtain the best agreement between the calculated and observed chemical shifts. The parameter values so obtained are then used in shielding calculations on the remaining molecules in Table 3.3.

It is perhaps not too surprising that the multi-centre shielding expressions should be sensitive to the values of the INDO parameters used to estimate the off-diagonal energy matrix elements. It is to be hoped that the optimized parameter values used in the shielding calculations will be suitable for the calculation of other molecular properties.

3.C.2 SOS Calculations

The first semi-empirical MO description of nuclear shielding, including all valence electrons, was presented by Pople.[31,32] This is still the most widely used approach in semi-empirical discussions of nuclear shielding.

Pople developed his SOS perturbation theory within the independent electron framework using the GIAOs originally introduced into the quantum theory of ring currents by London.[33,34] The approximations implicit in this model include setting all two-electron terms to zero and neglecting all two-centre overlap integrals. Consequently all two-centre terms which may be reduced to overlap-type integrals, e.g. $\langle \chi_\mu | L | \chi_\nu \rangle$, are systematically set to zero.

In addition to one-centre terms, the only remaining non-zero expressions are those for which both of the orbitals, χ_μ and χ_ν, are on another atomic centre, B. These expressions give rise to the non-local terms in the shielding expression; these terms are then analogous to equations (3.44) and (3.45) in the FPT procedure. All three-centre and higher terms are ignored in People's model. A further limitation arises from the use of a truncated valence orbital basis set with the result that the summations over MOs in equations (3.52) and (3.53) are truncated.

For atoms with only s and p valence electrons the expressions used for the local and non-local shielding contributions are as follows:

$$\sigma^d_{A\alpha\beta}(\text{loc.}) = \frac{\mu_0 e^2}{8\pi m} \sum_\mu^A P_{\mu\mu}\langle\chi_\mu|(\mathbf{r}_\mu\cdot\mathbf{r}_A\delta_{\alpha\beta} - r_{\mu\alpha}\cdot r_{A\beta})r_A^{-3}|\chi_\mu\rangle, \qquad (3.50)$$

$$\sigma^d_{A\alpha\beta}(\text{non-loc.}) = \frac{\mu_0 e^2}{16\pi m} \sum_{B\neq A}\sum_\gamma$$

$$\times \left[\sum_\lambda^B P_{\lambda\lambda}\langle\chi_\lambda|(\mathbf{r}_\lambda\cdot\mathbf{r}_A\delta_{\alpha\beta} - r_{\lambda\alpha}\cdot r_{A\beta})r_A^{-3}|\chi_\lambda\rangle\right.$$

$$\left. \times (R_B^2\delta_{\alpha\beta} - 3R_{B\gamma}R_{B\beta})R_B^{-5}\right], \qquad (3.51)$$

$$\sigma^p_{A\alpha\beta}(\text{loc.}) = -\frac{\mu_0 e^2 \hbar^2}{2\pi m^2} \sum_j^{\text{occ.}}\sum_k^{\text{unocc.}}(E_k - E_j)^{-1}\langle r^{-3}\rangle_{pA}$$

$$\times \sum_{\mu<\nu}^A (C_{\mu j}C_{\nu k} - C_{\nu j}C_{\mu k})\langle\chi_\mu|L_\alpha|\chi_\nu\rangle$$

$$\times \sum_{\lambda<\sigma}^B\sum^B (C_{\lambda j}C_{\sigma k} - C_{\sigma j}C_{\lambda k})\langle\chi_\lambda|L_\beta|\chi_\sigma\rangle, \qquad (3.52)$$

$$\sigma^p_{A\alpha\beta}(\text{non-loc.}) = -\frac{\mu_0 e^2 \hbar^2}{4\pi m^2} \sum_{B\neq A}\sum_j\left[\sum_j^{\text{occ.}}\sum_k^{\text{unocc.}}(E_k - E_j)^{-1}\right.$$

$$\times \sum_{\mu<\nu}^B (C_{\mu j}C_{\nu k} - C_{\nu j}C_{\mu k})\langle\chi_\mu|L_\alpha|\chi_\nu\rangle R_B^{-5}$$

$$\times (R_B^2\delta_{\alpha\beta} - 3R_{Bj}R_{B\beta})$$

$$\left. \times \sum_{\lambda<\sigma}^B\sum^B (C_{\lambda j}C_{\sigma k} - C_{\sigma j}C_{\lambda k})\langle\chi_\lambda|L_\gamma|\chi_\sigma\rangle\right], \qquad (3.53)$$

where R_B is the distance between nuclei A and B, j and k refer to occupied and unoccupied MOs, respectively, with energies of E_j and E_k, $(E_k - E_j)$ is the singlet excitation energy from j to k, given by

$$E_k - E_j = \varepsilon_k - \varepsilon_j - J_{jk} + 2K_{jk}, \qquad (3.54)$$

where ε_k and ε_j are the eigenvalues of the unperturbed molecule and J_{jk} and K_{jk} are the molecular Coulomb and exchange integrals, respectively.

The SOS formulation used to obtain equations (3.50) and (3.52) renders these expressions similar to equations (3.24) and (3.25) derived by applying the LCAO-SCF procedure to Ramsey's shielding model. However, the similarities are rather superficial in that Ramsey's expressions are molecular properties whereas Pople's model gives rise to localized shielding contributions. This is exemplified by the fact that all of the integrals in equations (3.50)–(3.53) are one-centre expressions, which is a consequence of the independent electron formulation of the shielding problem. The local terms, equations (3.50) and (3.52), incorporate the interactions expressed by equations (3.42), (3.43), (3.46) and (3.47).

Equations (3.51) and (3.53) for the non-local terms are produced by assuming that the field at nucleus A, due to the effect of electrons on atom B, may be approximated by replacing the induced moment of B by a point dipole.[35] From the presence of the matrix elements $\langle \chi_\mu | L_\alpha | \chi_\nu \rangle$ and $\langle \chi_\lambda | L_j | \chi_\sigma \rangle$ in equation (3.53) it is clear that non-local paramagnetic shielding contributions will only occur if both atoms A and B have valence p electrons.

The results shown in Table 3.4 indicate that the non-local diamagnetic

Table 3.4 Some values of σ^d(loc.), σ^d(non-loc.), σ^p(loc.) and σ^p(non-loc.) in parts per million, obtained from INDO/S parameterized SOS calculations

Molecule	Nucleus	σ^d(loc.)	σ^d(non-loc.)	σ^p(loc.)	σ^p(non-loc.)
$(CH_3)_2\overset{*}{C}O$	$\overset{*}{C}$	257.60	−0.12	−212.14	4.99
	O	398.47	0.10	−580.07	2.42
CO_2	C	255.16	−0.25	−126.79	−6.94
	O	397.32	0.12	−269.51	−3.55
HCN	C	259.07	0.11	−155.35	−5.88
	N	326.70	0.06	−301.30	−5.04
N_2	N	324.39	0.24	−382.84	−8.62
C_2H_2	C	260.62	−0.04	−144.18	−4.36
C_2H_4	C	260.41	0.01	−199.26	−0.26
C_6H_6	C	260.12	−0.01	−187.42	1.19

term is usually negligible, whereas the corresponding paramagnetic term can be important when multiple bonding occurs. A qualitatively similar situation exists with regard to the semi-empirical FPT results given in Table 3.3. However, the smaller non-local shielding contributions produced by Pople's expressions indicate one of the more serious shortcomings of this model. Notwithstanding this, Pople's expressions for the local diamagnetic and paramagnetic shielding terms are widely encountered and have been used to produce much chemically interesting information.[7]

By taking only a minimum basis set of atomic functions, χ, and rotationally averaging the expressions for the local shielding terms, equations (3.50) and (3.52) become

$$\sigma_A^d(\text{loc.}) = \frac{\mu_0 e^2}{12\pi m} \sum_\mu^A P_{\mu\mu} \langle \chi_\mu | r^{-1} | \chi_\mu \rangle \tag{3.55}$$

and

$$\sigma_A^p(\text{loc.}) = -\frac{\mu_0 e^2 \hbar^2}{6\pi m^2} \langle r^{-3} \rangle_{pA} \sum_j^{\text{occ.}} \sum_k^{\text{unocc.}} (E_k - E_j)^{-1}$$

$$\times \bigg[C_{y,Aj} C_{z,Ak} - C_{zAj} C_{yAk}) \sum_B (C_{y,Bj} C_{z,Bk} - C_{z,Bj} C_{y,Bk})$$

$$\times (C_{z,Aj} C_{x,Ak} - C_{x,Aj} C_{z,Ak}) \sum_B (C_{z,Bj} C_{x,Bk} - C_{x,Bj} C_{z,Bk})$$

$$\times (C_{x,Aj} C_{y,Ak} - C_{y,Aj} C_{x,Ak}) \sum_B (C_{x,Bj} C_{y,Bk} - C_{y,Bj} C_{x,Bk}) \bigg], \tag{3.56}$$

where $C_{x,Aj}$ is the LCAO coefficient of the p_x orbital on atom A in MO j etc.

It is clear from equation (3.56) that $\sigma_A^p(\text{loc.})$ is zero unless both atoms A and B possess p valence electrons. Thus hydrogen nuclei have no paramagnetic shielding contributions, as noted in Section 3.B for Ramsey's model. From the ordering of the LCAO coefficients in equation (3.56) it follows that all singlet electronic excitations, with the exception of $\pi \to \pi^*$, may be considered to contribute to the local paramagnetic shielding term. Thus all contributions arise from charge circulation effects and these are likely to be larger for systems having low-lying states localized on A.

The matrix element in equation (3.55) is usually evaluated as

$$\langle \chi_\mu | r^{-1} | \chi_\mu \rangle_A = Z_\mu / n^2 a_0 \tag{3.57}$$

and the expectation value for the mean inverse cube of the radius of the p orbitals on atom A, with primary quantum number n, is similarly obtained from

$$\langle r^{-3} \rangle_{pA} = \frac{1}{3} \left(\frac{Z_{np}}{n a_0} \right)^3 \tag{3.58}$$

where Z_μ is the effective nuclear charge for χ_μ and a_0 is the Bohr radius.

Values of Z_μ may be obtained from Slater's rules.[36] From these rules 2s and 2p orbitals on first row atoms have the same effective nuclear charge given by

$$Z_{2s} = Z_{2p} = Z^0 + 0.35 q_A^{net}, \qquad (3.59)$$

where Z^0 is the effective charge for the free atom and q_A^{net} is the net charge on atom A.

Equation (3.56) may be simplified by means of the average excitation energy, ΔE, approximation which is discussed with regard to equation (3.15) of Ramsey's model. Since overlap is neglected at the CNDO/INDO level of approximation it follows that

$$\sum_j^{occ.} C_{\mu j} C_{\lambda j} + \sum_k^{unocc.} C_{\mu k} C_{\lambda k} = \delta_{\mu\nu}. \qquad (3.60)$$

By incorporating equation (3.60) and the ΔE approximation, equation (3.56) becomes

$$\sigma_A^p(\text{loc.}) = -\frac{\mu_0 e^2 \hbar^2}{8\pi m^2 \Delta E} \langle r^{-3} \rangle_{pA} \sum_B Q_{AB}, \qquad (3.61)$$

where the summation over B includes atom A and all the other atoms in the molecule. The bond-order, charge-density terms, Q_{AB}, are given by

$$Q_{AB} = \tfrac{4}{3} \delta_{AB}(P_{x_A x_B} + P_{y_A y_B} + P_{z_A z_B})$$
$$- \tfrac{2}{3}(P_{x_A x_B} P_{y_A y_B} + P_{x_A x_B} P_{z_A z_B} + P_{y_A y_B} P_{z_A z_B})$$
$$+ \tfrac{2}{3}(P_{x_A y_B} P_{x_B y_A} + P_{x_A z_B} P_{x_B z_A} + P_{y_A z_B} P_{y_B z_A}). \qquad (3.62)$$

The presence of the Kronecker delta function, δ_{AB}, implies that the first term in equation (3.62) corresponds to the p electron charge density on atom A; the remaining two terms represent the multiple-bond contributions to the shielding tensor. This is perhaps best illustrated by considering the case of A and B being neighbours with the x axis chosen as the bond direction. Under these circumstances $P_{x_A x_B}$ is the bond order for the p orbitals forming the σ bond and $P_{y_A y_B}$ and $P_{z_A z_B}$ are the π bond orders. The final expression in equation (3.62) represents multiple-bond cross terms and it is usually rather small. It is clear that neighbouring atoms contribute to Q_{AB} when both σ and π bonding occurs between atoms A and B.

The most commonly encountered approach to calculating chemical shifts is that based upon equations (3.55) and (3.61). Numerous semi-empirical MO methods have been used to evaluate Q_{AB}. However, this procedure is of limited theoretical appeal in that it involves choosing a value for ΔE for which there is no *a priori* procedure. A value of 10 eV is often used for calculations of ^{13}C chemical shifts. While this may be a reasonable choice for saturated molecules it is likely to be too large to reproduce the substantial ^{13}C deshielding experienced in conjugated molecules.

Table 3.5 Comparison of some INDO/S-SOS calculated ^{13}C shielding results with experimental data (in parts per million)

Molecule	σ^d(loc.)	σ^p(loc.)	σ^p(non-loc.)	σ^{total}	$\sigma_{observed}$
$CH_2\!=\!CH_2$	260.41	−199.26	−0.26	60.89	74.0
$\diagdown\!\!=\!*$	260.62	−190.61	−0.13	69.88	81.4
$\diagup\!\!\!\diagdown\!=\!*$	260.81	−190.87	1.43	71.39	87.0
$\diagdown\!\!=\!\!_*$	260.12	−195.22	−0.13	64.77	$\begin{cases}63.7\\61.1\end{cases}$
$\diagdown\!\!=\!\!\diagup_*$	260.32	−194.24	−0.22	65.87	$\begin{cases}72.5\\73.5\end{cases}$
benzene	260.12	−187.42	1.19	73.88	68.1
$\diagup\!\!\!{}^*\!\!=\!\!$	259.84	−199.36	1.39	61.86	55.6
$\diagup\!\!\!\diagdown\!\!=\!\!\diagup^*\!\diagdown$	260.23	−197.05	−0.28	62.89	73.6
$CH_2\!=\!\overset{*}{C}\!=\!CH_2$	259.63	−274.44	−0.59	−15.41	−16.2
$CH_3\!-\!N\!\equiv\!\overset{*}{C}$	260.39	−188.09	−4.47	67.83	38.1
$CH_3\!-\!\overset{*}{C}\!\equiv\!N$	258.85	−158.55	−5.70	94.60	79.6
$HC\!\equiv\!CH$	260.67	−134.45	−4.15	122.08	120.0

In addition to the charge density terms appearing in equation (3.62), it is apparent from equations (3.58) and (3.59) that $\langle r^{-3}\rangle_{p_A}$ depends primarily on the charge density associated with atom A. Thus if for a given series of molecules the values of ΔE and the bond-order terms either remain constant or produce cancelling changes, then from equations (3.61) and (3.62) it follows that chemical shift trends can be dependent upon changes in charge density. Although such a correlation may be of some use in predicting possible chemical shifts for other members of the series of molecules considered, it does not provide much insight into the chemically interesting processes which give rise to the measured shieldings.

A particular advantage of semi-empirical calculations is the ease with which the results can be given a physical interpretation. SOS shielding calculations employing equations (3.55) and (3.56) can provide interesting results. In order to obtain a reasonable description of the electronic excited states, required for the evaluation of equation (3.56), parameter sets specifically dealing with these states should be incorporated into the shielding calculations.[7] Thus, as shown in Table 3.5, calculations using INDO/S parameters can provide a satisfactory account of the observed nuclear shielding differences. Analysis of the calculated data reveals the relative importance of the various terms, in equation (3.56), in determining the nuclear shielding.

Although the chemical shift is a scalar quantity it is important to appreciate that $\underline{\sigma}$ is a second rank tensor, as implied by equation (2.138). Thus, as mentioned in Chapter 1, $\underline{\sigma}$ may in principle be anisotropic. Its anisotropy may be measured by experiments on aligned molecules, such as those found in the solid state or in liquid crystal media. As solid state NMR becomes more widely practised it is anticipated that shielding anisotropies will be more widely reported.

Calculations of shielding anisotropies may be readily performed by the procedures outlined in this chapter, with the exception of the AEE approach. Since the anisotropy is the difference between diagonal shielding tensor components, errors arising from the necessary approximations inherent to semi-empirical MO calculations may reasonably be expected to be absent from estimates of the shielding anisotropy. This has been demonstrated to be the case for a number of first row nuclei.[7]

3.D RING CURRENTS

The interatomic shielding terms included in equation (3.40) encompass the hypothetical ring currents. Although these currents are not physically observable, they manifest themselves in proton NMR by the deshielding observed in the spectra of (poly)cyclic conjugated systems. Conversely, the cyclophane protons lying above a benzene ring have their shielding increased by the ring currents.

Shielding changes of a few parts per million are thought to arise from ring currents. Various classical and quantum mechanical theories have been proposed to account for these shielding effects.[8]

The classical point dipole model for estimating shielding influences of ring currents was developed by Pople.[37] In this procedure the effect on the benzene protons is assumed to be equivalent to that of a point dipole at the centre of a ring. The magnitude, M, of the dipole is given by

$$M = IA, \qquad (3.63)$$

where A is the area enclosed by the current loop and I is the ring current which may be deduced from

$$I = -3e^2 B/2\pi m. \qquad (3.64)$$

The direction of the applied magnetic field, B, is considered to be perpendicular to the molecular plane. If the six benzene π electrons are understood to flow along a circle of radius a, then

$$A = \pi a^2. \qquad (3.65)$$

A secondary field, B', is produced at the ring protons, which are at a distance R from the centre of the ring, where B and B' are collinear. A value for B' may be obtained from

$$B' = \mu_0 M/4\pi R^3. \qquad (3.66)$$

If B lies in the plane of the ring then the ring current is not operative; thus the rotationally averaged contribution from the secondary field to the proton shielding is due to $B'/3$.

The ring current shielding effect, $\Delta\sigma$, is often expressed by the ratio of the secondary field to the applied field at the nucleus of interest. In the case of benzene protons,

$$\Delta\sigma = -\mu_0 e^2 a^2/8\pi m R^3. \qquad (3.67)$$

The negative sign in equation (3.67) implies that proton deshielding is produced.

Within the framework of the classical point dipole model the current in each ring of a polycyclic molecule is taken to be equal to that in benzene, thus

$$\Delta\sigma = -\frac{\mu_0 e^2 a^2}{8\pi m} \sum \frac{1}{R_i^3}, \qquad (3.68)$$

where R_i is the separation of the proton concerned from the centre of the ith ring. This simple classical model accounts qualitatively for the proton deshielding observed for a restricted range of conjugated polycyclic hydrocarbons.[8] A slight improvement in the agreement between the calculated and observed ring current effects involves acknowledging that the currents will not flow in the molecular plane but in loops at a distance of 0.064 nm above and below the plane. This distance is chosen in order to mimic the proton deshielding of 1.5 p.p.m. found in benzene.

London[33,34] provided the basis for a quantum mechanical theory of ring current effects. In this approach the effective one-electron Hamiltonian, $\mathcal{H}$, is given by

$$\mathcal{H} = \frac{1}{2m}(p + eA)^2 + V, \tag{3.69}$$

where V is the potential energy term, p is the momentum of an electron in the absence of an applied magnetic field and eA is the momentum arising from the introduction of the field. Thus $\mathcal{H}$ gives rise to a set of secular equations which are modified to include the effect of the applied field.

The introduction of the GIAOs, defined by equation (3.36), gives rise to the matrix elements, $\mathcal{H}_{\mu\nu}$, where

$$\mathcal{H}_{\mu\nu} = \int \left(\exp\left[\frac{ie}{\hbar}(\mathbf{A}_\mu - \mathbf{A}_\nu)\cdot\mathbf{r}\right] \chi_\mu^* \left[\frac{1}{2m}(p + e(\mathbf{A} - \mathbf{A}_\nu)^2 + V\right]\chi_\nu \right) d\tau, \tag{3.70}$$

where $\mathbf{A}_\mu$ is defined by equation (3.37) and $\mathbf{r}$ is the position vector of an electron referred to an arbitrary origin. The corresponding overlap matrix elements, $S_{\mu\nu}$, are given by

$$S_{\mu\nu} = \int \left(\exp\left[\frac{ie}{\hbar}(\mathbf{A}_\mu - \mathbf{A}_\nu)\cdot\mathbf{r}\right] \chi_\mu^*\chi_\nu \right) d\tau. \tag{3.71}$$

In order to reduce equations (3.70) and (3.71) to a readily calculable form some assumptions are made which relate the matrix elements to those applicable in the absence of the applied field.

First, it is assumed that the exponential factors are unity when $\mu = \nu$. The residue of equation (3.70) describes the effect of the local diamagnetic circulation $e(\mathbf{A} - \mathbf{A}_\nu)$ on the energy of the atomic orbital χ_ν.

In dealing with interatomic currents, local terms are considered to be unimportant. Consequently, when the centres containing χ_μ and χ_ν are bonded, the contribution of $e(\mathbf{A} - \mathbf{A}_\nu)$ to the energy is ignored. Within the essence of Hückel theory, $\mathcal{H}_{\mu\nu} = 0$ if χ_μ and χ_ν are on centres which are not bonded and $\mathcal{H}_{\mu\mu} = \alpha$, which is customarily taken as the arbitrary zero of energy.

The second assumption which is made applies to directly bonded atoms; in this case the electron position vector, **r**, is replaced by the value it would take at the midpoint of the bond between the atoms, namely

$$\mathbf{r} = \tfrac{1}{2}(\mathbf{R}_\mu + \mathbf{R}_v), \tag{3.72}$$

where $\mathbf{R}_\mu$ and $\mathbf{R}_v$ are the position vectors of the atoms containing χ_μ and χ_v respectively.

As a result of this approximation the exponential factor in equations (3.70) and (3.71) is constant, for a given χ_μ and χ_v, and may be taken outside the integration operation. Hence,

$$\mathcal{H}_{\mu v} = \left[\exp\left(\frac{ie}{2\hbar}(\mathbf{A}_\mu - \mathbf{A}_v)\cdot(\mathbf{R}_\mu + \mathbf{R}_v)\right)\right]$$
$$\times \int \chi_\mu^* \left(\frac{p^2}{2m} + V\right) \chi_v \, d\tau \tag{3.73}$$

and

$$S_{\mu v} = \left[\exp\left(\frac{ie}{2\hbar}(\mathbf{A}_\mu - \mathbf{A}_v)\cdot(\mathbf{R}_\mu + \mathbf{R}_v)\right)\right] \int \chi_\mu^* \chi_v \, d\tau. \tag{3.74}$$

The introduction of the usual Hückel approximations to equations (3.73) and (3.74) lead to solutions of the secular equations as described in Chapter 2.

By analogy with equation (3.63) the induced π magnetic moment, M^π, opposing the applied field is given by,

$$M^\pi = I_i A_i, \tag{3.75}$$

where I_i is the induced line current around the ith ring of a conjugated system and A_i is the corresponding algebraic area. Now

$$M^\pi = -\frac{\partial E^\pi}{\partial B}, \tag{3.76}$$

where E^π is the total π electron energy in the presence of the applied field, B. The value of I_i is given by

$$I_i = -\frac{2e}{h} \sum_{k=1}^{\text{occ.}} \frac{\partial E_k}{\partial f_i} \tag{3.77}$$

since for a closed shell system

$$E^\pi = 2 \sum_{k=1}^{\text{occ.}} E_k \tag{3.78}$$

and

$$f_i = eBA_i/h. \tag{3.79}$$

Thus, from equation (3.77), the ring current intensity is expressed for a general, planar, polycyclic structure in an external magnetic field as a function of the molecular geometry, the strength of the applied field and the Hückel molecular energies.

McWeeny[38] has demonstrated that the ring current shielding effect may be expressed as

$$\Delta\sigma = \text{constant} \times \sum_i I_i[-K(\mathbf{r}_i)] \qquad (3.80)$$

where I_i may be calculated from geometrical and Hückel MO quantities, as demonstrated by equations (3.77)–(3.79), and $K(\mathbf{r}_i)$ is a geometrical factor which depends upon the position in space of the ith ring relative to the origin:

$$K(\mathbf{r}_i) = \sum_{\substack{\text{around} \\ \text{ring } i}} A_{gh} K_{gh}. \qquad (3.81)$$

In equation (3.81) A_{gh} is the signed area of the triangle formed by the origin and the atoms at the extremities of the g—h bond. The summation is taken around the ring whose centre is related to the origin by the position vector $\mathbf{r}_i$ and

$$K_{gh} = \frac{1}{r_g^3} + \frac{1}{r_h^3}, \qquad (3.82)$$

where r_g refers to the length of the vector relating the nucleus g of the conjugated system to the origin in terms of the length of a benzene carbon–carbon bond.

The shielding effect of a planar conjugated polycyclic structure is thus provided with a quantum mechanical foundation by means of equation (3.80). From this equation the sigma ratio, which compares the ring current deshielding effect on a given proton with that on a benzene proton, may be readily obtained:

$$\frac{\Delta\sigma}{\Delta\sigma_{\text{benzene}}} = \sum_i \frac{I_i}{I_{\text{benzene}}} \frac{K(\mathbf{r}_i)}{K_{\text{benzene}}}. \qquad (3.83)$$

In equation (3.83) I_i/I_{benzene} is the relative ring current intensity in the ith ring of a polycyclic molecule and $K(\mathbf{r}_i)/K_{\text{benzene}}$ is the relative geometric factor for the proton in question with respect to the ith ring of the conjugated system.

For a collection of 66 non-hindered protons in 16 *para*-condensed benzenoid hydrocarbons, it is found that calculating relative geometric factors by Pople's point dipole method and relative ring current intensities by McWeeny's procedure gives a more satisfactory account of the induced shifts than that obtained by more sophisticated models of ring current effects.[39] Included amongst the latter models are some based upon Hartree–Fock SCF theory.

Equation (3.80) is applicable to protons residing in the plane of a conjugated system. For protons in out-of-plane positions the signed area terms, A_{gh}, of equation (3.81) are replaced by their projections on the molecular plane, A_{gh}^{plane}; consequently $K(\mathbf{r}_i)$ in equation (3.80) is replaced by $K^1(\mathbf{r}_i)$, where

$$K^1(\mathbf{r}_i) = \sum_{\substack{\text{around} \\ \text{ring}\, i}} (A_{gh}^{plane}) K_{gh}. \qquad (3.84)$$

The modified version of equation (3.80) thus provides a measure of the intra- or intermolecular ring current effect of a conjugated ring system on the shielding of a proton in any position relative to it.

This approach has been adopted by Haigh and Mallion[40] in the production of a set of quantum mechanically based ring current shielding tables. As shown in Fig. 3.1, the calculations take account of the hexagonal structure of the benzene ring. The resulting shielding tables are expressed in terms of the cylindrical coordinates ρ, φ, z, defined in Fig. 3.1.

Some isoshielding lines for a plane perpendicular to the molecular plane and containing a line through one carbon atom and the centre of the ring, $\varphi = 0$, are shown in Fig. 3.2. The negative values of $\Delta\sigma$ for positions at the side of a ring signify a decrease in shielding, whereas the positive values for positions above a ring imply an increase in shielding.

By comparison with an extensive compilation of experimental NMR results it appears that the Haigh–Mallion tables are very reliable for nuclei in the deshielding zone, in and near the plane of the conjugated ring system. However, the tables tend to underestimate the extent of shielding experienced in the shielding cones above and below the plane of the ring system. This appears to be due to a breakdown of the London approximations in this region of space.

In contrast to this the classically derived Johnson–Bovey tables[41] overestimate the effects of the deshielding zone but appear to be more reliable in the shielding regions. Thus the complementary nature of the regions of space for which the shielding may be satisfactorily predicted, by these two sets of tables, may be judiciously exploited.

Both sets of tables have been employed in an analysis of ring current effects on the shielding of aliphatic protons in the basic pancreatic trypsin inhibitor.[42] The investigation indicates that ring currents are largely responsible for the conformation-dependent chemical shifts of the peripheral side chain protons. In contrast to this, ring currents appear to make only a small contribution to the conformation-dependent shifts of the backbone α- and amide-protons and the protons in the conjugated rings of the inhibitor. These conclusions are supported by other structural data and are useful for the assignment of individual proton signals.

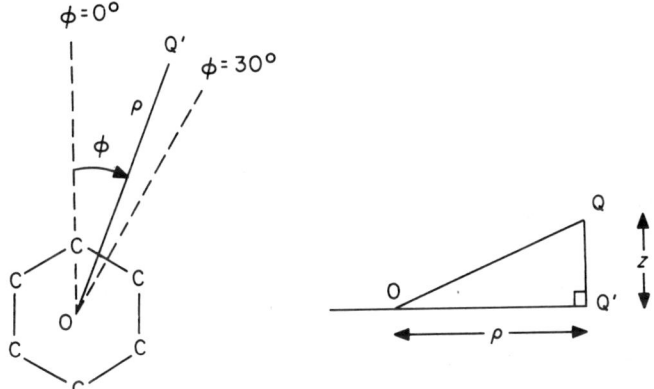

Fig. 3.1 Coordinate system used in ring current calculations.

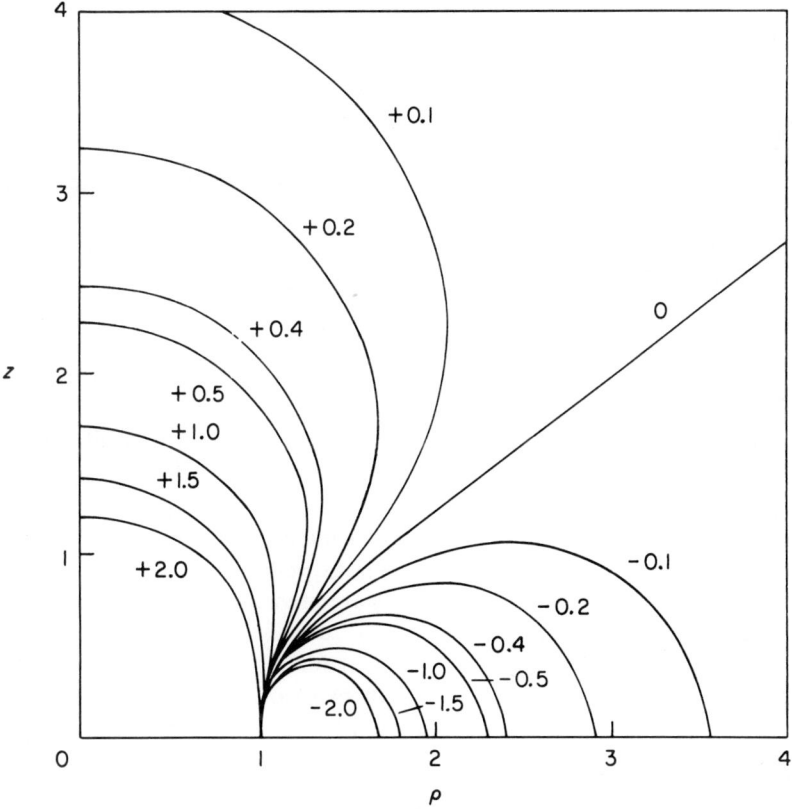

Fig. 3.2 Isoshielding contours for $\varphi = 0$. The $\Delta\sigma$ values are given in parts per million.

3.E MEDIUM EFFECTS ON NUCLEAR SHIELDING

Measured chemical shift variations between closely related molecules may arise not only from a change in chemical structure as such but also from possible differences in solute–solvent interactions. In addition, complications may be presented by an averaging over several conformations when a low barrier to conformational change exists.

It is possible to calculate shielding differences produced by conformational changes as well as those arising from solute–solvent interactions. The effects of solvents on solute molecules may be categorized into specific and non-specific classes. The specific interactions include hydrogen bonding, protonation, molecular association, ionic interactions and those producing aromatic solvent-induced shifts. From the theoretical point of view these may be treated by the supermolecule approach.[32]

Non-specific shielding influences can arise from bulk susceptibility effects, van der Waals' effects, neighbour susceptibility anisotropies and electric field effects. In essence the non-specific influences are produced by electronic interactions between the dipole moments of the solute and solvent molecules.[44]

A clear example of the influence of a specific interaction is afforded by the nitrogen shielding difference of 113 p.p.m. between pyridine and pyridinium ion. CNDO/S parameterized SOS calculations on these two species reveal that the shielding change is due to the removal of the $n \to \pi^*$ contribution to the local paramagnetic term upon protonation of the nitrogen lone pair electrons.[45] As a consequence of this the total nitrogen shielding for pyridine is smaller than that of the pyridinium ion.

The effects of hydrogen bonding on nuclear shielding are demonstrated by calculations on formamide surrounded by its first hydration shell, which is shown in Fig. 3.3.[46] The shielding calculation on this supermolecule has been performed at the *ab initio* level using GIAOs within the FPT framework.

The shielding results obtained for mono-, di- and tetrahydrated formamide are compared with those for the isolated molecule in Table 3.6. With the exception of the formyl proton, which is not specifically involved in hydrogen bonding, significant shielding variations are predicted for all of the nuclei of formamide upon hydration.

Tetrahydration produces a 7.4 p.p.m. reduction in the carbon shielding, which is in qualitative agreement with experimental data for hydrogen-bonded carbonyl functions. It appears that water molecules I and IV are those responsible for the largest carbon deshielding contributions. For the nitrogen nucleus the calculated deshielding of 10.2 p.p.m. agrees with experimental results on amides and peptides. In contrast to this the oxygen resonance is strongly shielded upon tetrahydration, the largest contributions

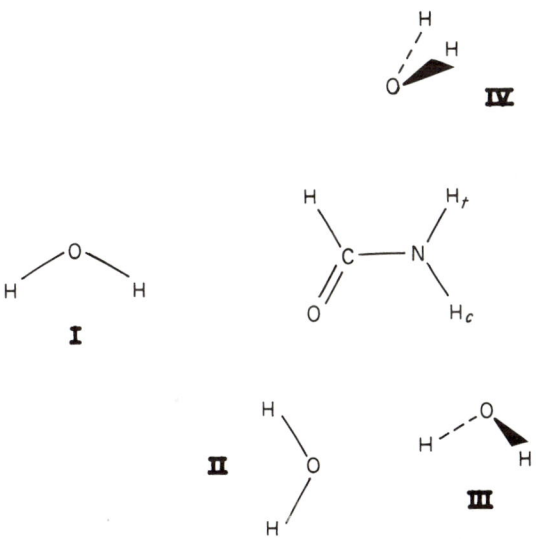

Fig. 3.3 Optimized geometrical arrangement of the first hydration shell of formamide.

coming from the directly hydrogen-bonded water molecules **I** and **II**. This observation is again in qualitative agreement with experiment.

The formation of the NH $\cdots$ O hydrogen bonds produces a deshielding of about 2 p.p.m. for the amide protons. Whereas the formation of the OH $\cdots$ O=C hydrogen bonds induces a further deshielding of about 0.4 p.p.m. For the isolated formamide molecule the shielding difference between the *cis* and *trans* protons, H_c and H_t, is calculated to be 0.12 p.p.m. This increases to 0.39 p.p.m. for the tetrahydrated molecule, in good agreement with the 0.37 p.p.m. reported at infinite dilution in water.

Table 3.6 Some calculated shielding data for formamide and some of its hydrated forms in parts per million

Nucleus	Formamide	Formamide dihydrate		Formamide tetrahydrate
		I + II	**III + IV**	
N	235.99	230.84	232.43	225.03
C	110.42	106.09	107.54	103.05
O	−216.50	−126.93	−181.01	−88.47
H(C)	23.27	23.18	23.33	23.22
$H(N)_c$	27.32	26.93	25.49	24.99
$H(N)_t$	27.20	26.79	25.08	24.60

The rapidly expanding field of high resolution NMR of solids has produced numerous examples of shielding differences both between a given molecule in the solid state and in solution and between molecules in the same unit cell.[47] Comparable effects are noted for small molecules absorbed on solid surfaces. It seems reasonable to anticipate that these shielding differences will be accounted for by calculations which include the effects of specific interactions.

Non-specific medium effects on NMR parameters have been estimated[48] by semi-empirical MO calculations involving the solvaton model.[44] In this model the following assumptions are made: (i) a solute molecule induces a number of charges (solvatons) in the solvent whose dielectric constant is ε; (ii) one solvaton is associated with each atom of the solute and its charge is equal in magnitude, but opposite in sign, to the net charge of the atom to which it is attached; (iii) no interactions occur between the solvatons.

Within this framework the solute–solvent interaction is incorporated into the Hamiltonian of the molecule, whose nuclear shielding is of interest, by means of $\mathcal{H}_{\text{solv.}}$, where

$$\mathcal{H}_{\text{solv.}} = \frac{\varepsilon - 1}{2\varepsilon} \left[\sum_{i=1}^{M} \sum_{s=1}^{N} \left(\frac{Q_s}{r_{si}} \right) - \sum_{s=1}^{N} \sum_{l=1}^{N} \left(\frac{Q_s Z_l}{r_{sl}} \right) \right] \quad (3.85)$$

for a molecule with M electrons and N nuclei, where Q_s is the induced solvaton charge, r_{si} and r_{sl} are the solvaton–electron and solvaton–nucleus separations respectively and Z_l is the nuclear charge.

The perturbation due to $\mathcal{H}_{\text{solv.}}$ results in a SOS local paramagnetic shielding term given by

$$\sigma^{\text{p}}(\text{loc.}) = -\left(\frac{\varepsilon - 1}{2\varepsilon} \right) \frac{\mu_0 e^2 \hbar^2}{3\pi m^2} \langle r^{-3} \rangle_{\text{p}} \sum_{j}^{\text{occ.}} \sum_{k}^{\text{unocc.}} \frac{C_{\mu k} C_{\nu k} - C_{\mu j} C_{\nu j}}{(E_k - E_j)^2}$$

$$\times \left\langle \chi_\mu \left| \sum_{i=1}^{M} \sum_{s=1}^{N} \left(\frac{Q_s}{r_{si}} \right) - \sum_{s=1}^{N} \sum_{l=1}^{N} \left(\frac{Q_s Z_l}{r_{sl}} \right) \right| \chi_\nu \right\rangle$$

$$\times [2(C_{yj}C_{yk}C_{zj}C_{zk} + C_{xj}C_{xk}C_{zj}C_{zk} + C_{xj}C_{xk}C_{yj}C_{yk})$$

$$- (C_{xj}C_{yk})^2 - (C_{yj}C_{xk})^2 - (C_{xj}C_{zk})^2$$

$$- (C_{zj}C_{xk})^2 - (C_{yj}C_{zk})^2 - (C_{zj}C_{yk})^2]. \quad (3.86)$$

Since changes in nuclear shielding are largely determined by those in $\sigma^{\text{p}}(\text{loc.})$, it follows from equation (3.86) that the solvaton model describes a variation in shielding which depends upon $(\varepsilon - 1)/2\varepsilon$. This is similar to that predicted by the reaction field theory.[49]

An example of the quantitative nature of the agreement between the observed and calculated nitrogen shielding changes of nitromethane is presented in Table 3.7. The calculated results are obtained from the INDO/S parameterized SOS procedure incorporating the solvaton model.[50] The choice of solvents is such as to minimize specific solute–solvent interactions.

The nitrogen shielding decreases as ε increases, which is consistent with the induced increase in the polarity of the nitromethane. As this occurs the positive charge on the nitrogen is enhanced, thus $\langle r^{-3}\rangle_p$ becomes larger. The resulting increase in σ^p(loc.) provides the observed nitrogen shielding decrease.

Table 3.7 Comparison of calculated and observed solvent effects on the nitrogen shielding of nitomethane, in parts per million

Solvent (0.30 M solutions)	Dielectric constant, ε, at 30 °C	Nitrogen shielding	
		Observed	Calculated
CCl_4	2.71	7.10 ± 0.11	8.8
Diethyl ether	4.79	3.91 ± 0.13	4.8
$CHCl_3$	5.07	3.79 ± 0.13	4.3
CH_2Br_2	6.78	3.41 ± 0.12	2.3
CH_2Cl_2	9.50	3.21 ± 0.12	1.3
Acetone	20.4	0.77 ± 0.10	0.0
CH_3CN	36.6	0.20 ± 0.13	−0.4
None	35.9	0.0000	−0.4
DMF	37.5	−0.69 ± 0.13	−0.5
DMSO	45.9	−2.01 ± 0.12	−0.6

The influence of intermolecular effects on nuclear shielding may also be described by means of the nuclear magnetic shielding density function which is represented by a surface in four-dimensional space. The shielding density function is analogous to the charge density; upon integration these functions yield the nuclear shielding and total number of electrons, respectively.

Shielding density maps of H, F^- and HF have been used to interpret shielding changes upon bond formation. Calculations of this type on more chemically interesting systems could provide the basis for an understanding of the shielding changes brought about by intermolecular effects.

3.F CONCLUSIONS

It is clear that protons, and other atoms with similar electronic configurations, form the exception rather than the rule in discussions of the molecular factors determining nuclear shielding. For the other NMR nuclei in the periodic table a few generalizations appear to be in order.

Ab initio procedures, using GIAOs and the FPT approach, appear to be the most satisfactory way of obtaining an understanding of the factors controlling the nuclear shielding of small molecules. However, it is important to appreciate that even if a complete basis set is used in a calculation which is pursued to the Hartree–Fock limit there are still defects in the solutions obtained, which arise from approximations used in the Hartree–Fock equations and in the perturbation treatment of the second-order molecular properties.

The Hartree–Fock equations suffer from two sources of error. The first arises from the fact that they are based on the Schrödinger equation, which normally employs operators which are independent of relativity effects. Both nuclear shielding and nuclear spin–spin coupling interactions are critically dependent upon electron motions close to the nuclei in question. The velocities of these inner electrons are dependent upon nuclear charge, thus for heavier nuclei a relativistic increase in the electronic mass has to be considered. As a result of this the radial functions of the s and p electrons are contracted; concomitant with this the effective nuclear charge for the d and f electrons decreases and thus their radial components increase in size. In addition, the atomic spin–orbital coupling interaction is proportional to the fourth power of the effective nuclear charge and thus becomes of greater significance for heavier nuclei. The result being that j–j, rather than Russel–Saunders, coupling of electronic angular momenta has to be considered.

These influences are combined in a relativistic MO approach to molecular problems and it seems very likely that they will play a very significant role in our understanding of the NMR parameters of heavier nuclei.

The second major shortcoming of the Hartree–Fock equations is that they do not include electron pair effects which give rise to correlation energies. Some compensation for this neglect can be made by including configuration interaction at the *ab initio* level.

When considering larger molecules, semi-empirical MO methods are currently employed in order to obtain results without the use of prohibitive amounts of computational effort. For calculations of nuclear shielding the SOS procedure with INDO/S parameters, or the FPT approach with INDO parameters, appear to be the most readily acceptable. Relativistically parameterized semi-empirical shielding calculations are confidently expected for heavier nuclei in the near future.

Finally, calculations based on the solvaton model show that the concept of a molecule in a vacuum is not too realistic for a consideration of the nuclear shielding of polar species. Similarly, the supermolecule approach demonstrates the importance of specific medium effects on nuclear shielding.

REFERENCES

1. N. F. Ramsey, *Phys. Rev.*, **78**, 699 (1950).
2. C. P. Slichter, "Principles of Magnetic Resonance", Springer-Verlag, Berlin (1978).
3. J. O. Hirschfelder, W. Byers-Brown and S. T. Epstein, *Adv. Quantum Chem.*, **1**, 255 (1964).
4. J. H. van Vleck, "The Theory of Electric and Magnetic Susceptibilities", Clarendon Press, Oxford (1932).
5. G. A. Webb, *in* "Annual Reports on NMR Spectroscopy", Vol. 6A (E. F. Mooney, ed.), Academic Press, London (1975), p. 1.
6. R. Ditchfield and P. D. Ellis, *in* "Topics in ^{13}C NMR Spectroscopy", Vol. 1 (G. C. Levy, ed.), John Wiley & Sons, New York (1974), p. 1.
7. K. A. K. Ebraheem and G. A. Webb, *in* "Progress in NMR Spectroscopy", Vol. 11 (J. W. Emsley, J. Feeney and L. H. Sutcliffe, eds), Pergamon Press, Oxford (1978), p. 149.
8. C. Haigh and R. Mallion, *in* "Progress in NMR Spectroscopy", Vol. 13 (J. W. Emsley, J. Feeney and L. H. Sutcliffe, eds), Pergamon Press, Oxford (1980), p. 303.
9. M. Karplus and H. J. Kolker, *J. Chem. Phys.*, **35**, 2235 (1961).
10. M. Karplus and H. J. Kolker, *J. Chem. Phys.*, **38**, 1263 (1963).
11. R. M. Stevens and W. N. Lipscomb, *J. Chem. Phys.*, **40**, 2238 (1964).
12. R. M. Stevens and W. N. Lipscomb, *J. Chem. Phys.*, **41**, 3710 (1964).
13. R. M. Stevens and W. N. Lipscomb, *J. Chem. Phys.*, **42**, 3660 (1965).
14. R. M. Stevens and W. N. Lipscomb, *J. Chem. Phys.*, **42**, 4302 (1965).
15. R. Ditchfield, D. P. Miller and J. A. Pople, *J. Chem. Phys.*, **53**, 613 (1970).
16. R. Ditchfield, D. P. Miller and J. A. Pople, *J. Chem. Phys.*, **54**, 4186 (1971).
17. H. Fukui, H. Yoshida and K. Miura, *J. Chem. Phys.*, **74**, 6988 (1981).
18. T. D. Gierke and W. H. Flygare, *J. Am. chem. Soc.*, **94**, 7277 (1972).
19. W. H. Flygare and J. Goodisman, *J. Chem. Phys.*, **49**, 3122 (1968).
20. G. Malli and C. Froese, *Int. J. Quantum Chem.*, **1s**, 95 (1967).
21. P. Lazzeretti and R. Zanasi, *J. Chem. Phys.*, **72**, 6768 (1980).
22. F. Keil and R. Ahlrichs, *J. Chem. Phys.*, **71**, 267 (1979).
23. J. Ridard, B. Levy and P. Millie, *Molec. Phys.*, **36**, 1025 (1978).
24. R. Ditchfield, *J. Chem. Phys.*, **56**, 5688 (1972).
25. R. Ditchfield, *Molec. Phys.*, **27**, 789 (1974).
26. F. R. Prado, C. Giessner-Prettre, B. Pullmann and J. P. Daudey, *J. Amer. chem. Soc.*, **101**, 1737 (1979).
27. J. A. Pople and D. L. Beveridge "Approximate Molecular Orbital Theory", McGraw-Hill, New York (1970).
28. A. R. Garber, P. D. Ellis, K. Seidman and K. Schade, *J. magn. Reson.*, **34**, 1 (1979).
29. P. D. Ellis, G. E. Maciel and J. W. McIver, *J. Am. chem. Soc.*, **96**, 4069 (1972).
30. P. D. Ellis, Y. C. Chou and P. A. Dobh, *J. magn. Reson.*, **39**, 529 (1980).
31. J. A. Pople, *J. Chem. Phys.*, **37**, 53 (1963).
32. J. A. Pople, *J. Chem. Phys.*, **37**, 60 (1963).
33. F. London, *J. Phys. Radium*, **8**, 397 (1937).
34. F. London, *J. Chem. Phys.*, **5**, 837 (1937).
35. J. A. Pople, *Proc. R. Soc. A*, **239**, 541 (1957).

36. J. C. Slater, *Phys. Rev.*, **36**, 57 (1930).
37. J. A. Pople, *J. Chem. Phys.*, **24**, 1111 (1956).
38. R. McWeeny, *Molec. Phys.*, **1**, 311 (1958).
39. R. B. Mallion, *J. Chem. Phys.*, **75**, 793 (1981).
40. C. W. Haigh and R. B. M. Mallion, *Org. magn. Reson.*, **4**, 203 (1972).
41. C. E. Johnson and F. A. Bovey, *J. Chem. Phys.*, **29**, 1012 (1958).
42. S. J. Perkins and K. Wüthrich, *Biochim. biophys. Acta*, **576**, 409 (1979).
43. G. A. Webb and M. Witanowski, *in* "Molecular Interactions", Vol. 5, in press.
44. I. Ando and G. A. Webb, *Org. magn. Reson.*, **15**, 111 (1981).
45. K. A. K. Ebraheem, G. A. Webb and M. Witanowski, *Org. magn. Reson.*, **11**, 27 (1978).
46. F. R. Prado, C. Giessner-Prettre, A. Pullman, J. F. Hinton, D. Harpool and K. Metz, *Theor. Chim. Acta*, **59**, 55 (1981).
47. R. E. Wasylishen and C. A. Fyfe, *in* "Annual Reports on NMR Spectroscopy", Vol. 12 (G. A. Webb, ed.), Academic Press, London (1982), p. 1.
48. H. A. Germer, *Theor. chim. Acta*, **34**, 145 (1974).
49. A. D. Buckingham, *Can. J. Chem.*, **38**, 300 (1960).
50. M. Witanowski, L. Stefaniak, B. Na Lamphun and G. A. Webb, *Org. magn. Reson.*, **16**, 57 (1981).
51. C. J. Jameson and A. D. Buckingham, *J. Chem. Phys.*, **73**, 5684 (1980).

4
Spin–Spin Couplings

Nuclear spin–spin couplings contain a wealth of information about molecular electronic structure. The couplings are sensitive to many aspects of molecular structure. Consequently, the general nature of the coupling mechanism is best considered through an understanding of the electronic structures of the molecules concerned.[1-5]

4.A GENERAL THEORETICAL BACKGROUND

After the discovery of indirect nuclear spin–spin coupling by Gutowsky and coworkers[6,7] and Hahn and Maxwell,[8] Ramsey and Purcell[9,10] were the first to provide a suitable theoretical interpretation for the mechanism of indirect spin–spin coupling. The theory of the coupling is based upon three types of electron-coupled interactions between the electrons and nuclei of the molecule concerned. As described briefly in Chapter 2, these are (i) a *contact* interaction between the electron and nuclear spins; (ii) a magnetic *dipolar* interaction between the electron and nuclear spins; (iii) an *orbital* interaction between the magnetic field produced by the orbital motion of the electrons and the nuclear magnetic dipole.

In most cases, the *contact* term,

$$\mathcal{H}^3 = \frac{4\mu_0\mu_B\hbar}{3} \sum_j \sum_N \gamma_N \, \delta(\mathbf{r}_{jN})(\mathbf{S}_j \cdot \mathbf{I}_N), \tag{4.1}$$

in the Hamiltonian provides the largest contribution to the total nuclear spin–spin interaction. The energy of interaction between nuclei N and N', $E_{NN'}$, is represented by second-order perturbation theory as

$$E_{NN'} = -\sum_{n(\neq 0)} \frac{1}{E_n - E_0} \langle \Psi_0 | \mathcal{H}' | \Psi_n \rangle \langle \Psi_n | \mathcal{H}' | \Psi_0 \rangle. \tag{4.2}$$

By substituting equation (4.1) into equation (4.2), we have

$$E_{NN'} = -\frac{16\mu_0^2\mu_B^2\hbar^2}{9}\gamma_N\gamma_{N'}\sum_{n(\neq 0)}\sum_k\sum_j\frac{1}{E_n - E_0}$$
$$\times \langle\Psi_0^0|\delta(\mathbf{r}_{kN})\mathbf{S}_k\mathbf{I}_N|\Psi_n^0\rangle\langle\Psi_n^0|\delta(\mathbf{r}_{jN'})\mathbf{S}_j\mathbf{I}_{N'}|\Psi_0^0\rangle, \quad (4.3)$$

where $E_n - E_0$ is the excitation energy from a ground spin singlet state to an excited spin triplet state, whose eigenfunctions $|\Psi_0^0\rangle$ and $|\Psi_n^0\rangle$, respectively, may be written as $|0\rangle$ and $|n\rangle$. Various authors have proposed that the nuclear spin–spin coupling may be interpreted in terms of an indirect interaction between two nuclei N and N' as

$$E_{NN'} = hJ_{NN'}\mathbf{I}_N\mathbf{I}_{N'}. \quad (4.4)$$

This is comparable to equation (4.3) and leads to the following expression for the contact contribution, $J_{NN'}^c$, to the spin–spin coupling constant $J_{NN'}$:

$$J_{NN'}^c = -\frac{8\mu_0^2\mu_B^2\hbar}{9\pi}\gamma_N\gamma_{N'}\sum_{n(\neq 0)}\sum_k\sum_j\frac{1}{E_n - E_0}$$
$$\times \langle 0|\delta(\mathbf{r}_{kN})\mathbf{S}_k|n\rangle\langle n|\delta(\mathbf{r}_{jN'})\mathbf{S}_j|0\rangle. \quad (4.5)$$

It is difficult to calculate the contact contribution to coupling constants using equation (4.5) because we have little knowledge of the exact wavefunctions of the excited states. An AEE approximation for all states has sometimes been applied to the calculation of spin–spin couplings. This approximation applied to equation (4.5) provides the following equation:

$$J_{NN'}^c = -\frac{8\mu_0^2\mu_B^2\hbar}{9\pi}\gamma_N\gamma_{N'}\frac{1}{\Delta E}\sum_{n(\neq 0)}\sum_k\sum_j$$
$$\times \langle 0|\delta(\mathbf{r}_{kN})\mathbf{S}_k|n\rangle\langle n|\delta(\mathbf{r}_{jN'})\mathbf{S}_j|0\rangle, \quad (4.6)$$

where ΔE is the spin singlet–triplet average excitation energy. The matrix sum rule of quantum mechanics is represented by

$$\sum_{n=0}|n\rangle\langle n| = 1, \quad (4.7)$$

which is often called the closure relation. If the closure relation is applied to equation (4.6), we have

$$J_{NN'}^c = -\frac{8\mu_0^2\mu_B^2\hbar}{9\pi}\gamma_N\gamma_{N'}\frac{1}{\Delta E}\sum_{n=0}\sum_k\sum_j\{\langle 0|\delta(\mathbf{r}_{kN})\mathbf{S}_k|n\rangle$$
$$\times \langle n|\delta(\mathbf{r}_{jN'})\mathbf{S}_j|0\rangle - \langle 0|\delta(\mathbf{r}_{kN})\mathbf{S}_k|0\rangle\langle 0|\delta(\mathbf{r}_{jN'})\mathbf{S}_j|0\rangle\}$$
$$= -\frac{8\mu_0^2\mu_B^2\hbar}{9\pi}\gamma_N\gamma_{N'}\frac{1}{\Delta E}\sum_k\sum_j\langle 0|\delta(\mathbf{r}_{kN})\delta(\mathbf{r}_{jN'})\mathbf{S}_k\mathbf{S}_j|0\rangle, \quad (4.8)$$

where the term

$$\langle 0| \sum_k \delta(\mathbf{r}_{kN})\mathbf{S}_k|0\rangle^2$$

is zero because the spin operator $\delta(\mathbf{r}_{kN})\mathbf{S}_k$ only mixes singlet and triplet states and the only states available are singlets in this instance.

In the case of spin–spin couplings involving hydrogen nuclei the dipolar and orbital interactions are usually neglected. However, they are very important in some examples of coupling not involving hydrogen nuclei such as nitrogen–carbon, fluorine–fluorine, carbon–fluorine and carbon–carbon couplings. In fluorine–fluorine couplings the non-contact interactions can predominate in magnitude over the contact interactions.

A magnetic *dipolar* interaction between the electronic and nuclear spins is given by

$$\mathcal{H}^2 = \frac{\mu_0 \mu_B \hbar}{2\pi} \sum_N \sum_k \gamma_N \{3(\mathbf{S}_k \cdot \mathbf{r}_{kN})(\mathbf{I}_N \cdot \mathbf{r}_{kN})r_{kN}^{-5} - \mathbf{S}_k \cdot \mathbf{I}_N r_{kN}^{-5}\}. \qquad (4.9)$$

By means of second-order perturbation theory, the above interaction gives the corresponding spin–spin coupling contribution as

$$J_{NN'}^d = -\frac{\mu_0^2 \mu_B^2 \hbar}{24\pi^3} \gamma_N \gamma_{N'} \sum_{n(\neq 0)} \sum_k \sum_j \frac{1}{E_n - E_0}$$
$$\times \langle 0|3(\mathbf{S}_k \cdot \mathbf{r}_{kN})\mathbf{r}_{kN} r_{kN}^{-5} - \mathbf{S}_k r_{kN}^{-3}|n\rangle$$
$$\times \langle n|3(\mathbf{S}_j \cdot \mathbf{r}_{jN'})\mathbf{r}_{jN'} r_{jN'}^{-5} - \mathbf{S}_j r_{jN'}^{-3}|0\rangle, \qquad (4.10)$$

where $E_n - E_0$ is the singlet–triplet excitation energy. Finally, the *orbital* interaction is divided into two terms: one is an orbital contribution involving a matrix element of $\mathcal{H}_1^1$ which only concerns the singlet ground state; the other is an orbital contribution involving a matrix element of $\mathcal{H}_2^1$ which mixes a singlet ground state with triplet excited states:

$$\mathcal{H}_1^1 = \frac{\mu_0^2 \mu_B^2 e \hbar}{16\pi^2} \sum_k \sum_N \sum_{N'} \gamma_N \gamma_{N'} (\mathbf{I}_N \cdot \mathbf{r}_{kN}/r_{kN}^3) \cdot (\mathbf{I}_{N'} \cdot \mathbf{r}_{kN'}/r_{kN'}^3), \qquad (4.11)$$

$$\mathcal{H}_2^1 = \frac{\mu_0 \mu_B \hbar}{2\pi i} \sum_k \sum_N \gamma_N (\mathbf{I}_N \cdot \mathbf{r}_{kN}/r_{kN}^3) \times \nabla_k$$

$$= \frac{\mu_0 \mu_B}{\pi} \sum_k \sum_N \gamma_N \mathbf{I} \cdot \mathbf{m}_{kN}^o / r_{kN}^3, \qquad (4.12)$$

where

$$\mathbf{m}_{kN}^o = m_{xk}^o \mathbf{i} + m_{yk}^o \mathbf{k} + m_{zk}^o \mathbf{j}. \qquad (4.13)$$

By means of second-order perturbation theory and selection of the terms dependent on $\mathbf{I}_N$ and $\mathbf{I}_{N'}$, we have

$$J^o_{NN'} = J^o_{NN'_1} + J^o_{NN'_2}, \tag{4.14}$$

where

$$J^o_{NN'_1} = \frac{\mu_0^2 e^2 \hbar}{48\pi^3 m} \gamma_N \gamma_{N'} \langle 0|\mathbf{r}_{kN}\mathbf{r}_{kN'} r_{kN}^{-3} r_{kN'}^{-3}|0\rangle, \tag{4.15}$$

$$J^o_{NN'_2} = -\frac{\mu_0^2 \mu_B^2 \hbar}{6\pi^3} \gamma_N \gamma_{N'} \sum_{n(\neq 0)} \sum_k \sum_j \frac{1}{E_n - E_0}$$

$$\times \langle 0|\mathbf{m}^o_{kN} r_{kN}^{-3}|n\rangle \langle n|\mathbf{m}^o_{jN'} r_{jN'}^{-3}|0\rangle. \tag{4.16}$$

The dipolar and orbital terms can be expressed by means of the AEE approximation rather than by equations (4.10) and (4.16) as in the case of the contact term.

Sometimes the electron-coupled interactions between nuclear spins are not isotropic. In this case we have to consider that the coupling is represented by a second-rank tensor $\mathbf{J}_{NN'}$ instead of the scalar $J_{NN'}$ because the matrix elements of equation (4.2) imply integration only over the coordinates of the electrons. As described by equation (2.146) the corresponding interaction energy is given by[8]

$$E_{NN'} = h\mathbf{I}_N \cdot \mathbf{J}_{NN'} \cdot \mathbf{I}_{N'}. \tag{4.17}$$

This term vanishes in molecules moving isotropically because the tensor $\mathbf{J}_{NN'}$ is traceless. Thus, by second-order perturbation theory we obtain the following three tensor expressions corresponding to equations (4.5), (4.10), (4.15) and (4.16) for the scalars which are normally considered in high resolution NMR spectroscopy:

$$\mathbf{J}^c_{NN'} = 3J^c_{NN'} - J^c_{NN'}\mathbf{E}, \tag{4.18}$$

$$\mathbf{J}^d_{NN'} = 3J^d_{NN'} - J^d_{NN'}\mathbf{E}, \tag{4.19}$$

$$\mathbf{J}^o_{NN'_1} = \frac{\mu_0^2 e^2 \hbar}{48\pi^3 m} \gamma_N \gamma_{N'} \langle 0| \sum [\mathbf{E}\mathbf{r}_{kN}\mathbf{r}_{kN'}$$

$$- \mathbf{r}_{kN}\mathbf{r}_{kN'}]r_{kN}^{-3} r_{kN'}^{-3}|0\rangle - J^o_{NN'_1}, \tag{4.20}$$

$$\mathbf{J}^o_{NN'_2} = 3J^o_{NN'_2} - J^o_{NN'_2}\mathbf{E}, \tag{4.21}$$

where $\mathbf{E}$ is the unit dyadic.

Ramsey[8] applied the above theory to the numerical calculation of the spin coupling of the molecule HD. The contact term is the major contributor in this case; it was estimated by means of the AEE approximation using equation (4.8) as follows:

$$J_{HD}^c = -\frac{8\mu_0^2\mu_B^2\hbar}{9\pi}\gamma_H\gamma_D\frac{1}{\Delta E}$$
$$\times \langle 0_e| \delta(\mathbf{r}_H) \delta(\mathbf{r}_D)|0_e\rangle\langle 0_n|\mathbf{S}_1\mathbf{S}_2|0_n\rangle \quad (4.22)$$

where the electronic orbital and nuclear spin wavefunctions are separated in the ground state with 0_e designating the former and 0_n the latter. The matrix element of $\langle 0_e| \delta(\mathbf{r}_H) \delta(\mathbf{r}_D)|0_e\rangle$ refers to the electron density on the coupled proton and deuteron. In equation (4.22), the electron spin vector product $\mathbf{S}_1 \cdot \mathbf{S}_2$, upon expansion, gives

$$\mathbf{S}_1 \cdot \mathbf{S}_2 = S_{1x} \cdot S_{2x} + S_{1y} \cdot S_{2y} + S_{1z} \cdot S_{2z}$$
$$= \tfrac{1}{2}(S_{1+}S_{2-} + S_{1-}S_{2+}) + S_{1z} \cdot S_{2z}, \quad (4.23)$$

where

$$S_{1+} = (S_{1x} + iS_{1y}), \quad (4.24)$$
$$S_{1-} = (S_{1x} - iS_{1y}). \quad (4.25)$$

The relations $S_z\alpha = \tfrac{1}{2}\alpha$, $S_z\beta = -\tfrac{1}{2}\beta$, $S_+\alpha = S_-\beta = 0$, $S_+\beta = \alpha$ and $S_-\alpha = \beta$ lead to

$$\mathbf{S}_1 \cdot \mathbf{S}_2 = \tfrac{1}{4}(2P_{12} - 1); \quad (4.26)$$

where P_{12} is a permutation operator which interchanges the spin coordinates of electrons 1 and 2. Following this relation, we have

$$\langle 0_n|\mathbf{S}_1 \cdot \mathbf{S}_2|0_n\rangle = \langle 0_n|\tfrac{1}{4}(2P_{12} - 1)|0_n\rangle = -\tfrac{3}{4}. \quad (4.27)$$

Thus, equation (4.22) becomes

$$J_{HD}^c = \frac{2\mu_0^2\mu_B^2\hbar}{3\pi}\gamma_H\gamma_D\frac{1}{\Delta E}\langle 0_e| \delta(\mathbf{r}_H) \delta(\mathbf{r}_D)|0_e\rangle$$
$$= \frac{2\mu_0^2\mu_B^2\hbar}{3\pi}\gamma_H\gamma_D\frac{1}{\Delta E}S_H(0)^2S_D(0)^2, \quad (4.28)$$

where

$$S_H(0)^2S_D(0)^2 = \langle 0_e| \delta(\mathbf{r}_H) \delta(\mathbf{r}_D)|0_e\rangle = 0.06/a_0^6 \quad (4.29)$$

in which $a_0 = \hbar^2/me^2$. Thus, using these expressions, we have

$$J_{HD} = 758.9/\Delta E \quad \text{(Hz)} \quad (4.30)$$

If ΔE is given a somewhat large value such as 19.0 eV for the singlet–triplet excitation energy, the value of J_{HD} calculated is 40 Hz, which is near the experimental value of 43 Hz.[11]

4.B MO APPROACHES

In early theoretical works, the calculation of spin–spin couplings was performed by two approaches, namely the MO and VB methods. The mathematical difficulties encountered, and problems associated, with the choice of a suitable algorithm necessary for computing the ground state VB wavefunction renders the VB method less satisfactory than the MO approach. For these reasons, current calculations of spin–spin couplings have been mainly carried out by means of MO expressions such as those arising from SOS, FPT and self-consistent perturbation (SCP) theory. In this section we are concerned with these three perturbation approaches to the calculation of spin–spin couplings.

4.B.1 SOS Method

The SOS method was developed by Pople and Santry[12] for the contact, orbital and dipolar contributions to spin–spin coupling. Nuclear spin–spin coupling involves a knowledge of the electronic energies of all excited triplet states mixing with the singlet ground state. There are four excitations of a single electron from an occupied MO φ_i into an unoccupied one φ_j to be considered. By means of the representations given in Section 2.B.1 the corresponding one-singlet and three-triplet wavefunctions can be written as

$$^{1}\Psi_{i\to j} = \frac{1}{\sqrt{2}}\{|\varphi_1\bar{\varphi}_2 \cdots \varphi_i\bar{\varphi}_j \cdots| + |\varphi_1\bar{\varphi}_2 \cdots \bar{\varphi}_j\varphi_i \cdots|\} \tag{4.31}$$

$$^{3}\Psi_{i\to j} = \begin{cases} |\varphi_1\bar{\varphi}_2 \cdots \bar{\varphi}_i\varphi_j \cdots|, & (4.32) \\ (1/\sqrt{2})\{|\varphi_1\bar{\varphi}_2 \cdots \varphi_i\bar{\varphi}_j \cdots| - |\varphi_1\bar{\varphi}_2 \cdots \bar{\varphi}_j\varphi_i \cdots|\}, & (4.33) \\ |\varphi_1\bar{\varphi}_2 \cdots \varphi_i\varphi_j \cdots|. & (4.34) \end{cases}$$

The excitation energies corresponding to $^{1}\Psi_{i\to j}$ and $^{3}\Psi_{i\to j}$ are expressed as $\Delta^{1}E_{i\to j}$ and $\Delta^{3}E_{i\to j}$, respectively.

The expression for the contact term (equation (4.5)) is rewritten using equations (4.31)–(4.34) as

$$J^{c}_{NN'} = -\frac{8\mu_0^2\mu_B^2 \hbar}{9\pi}\gamma_N\gamma_{N'}\sum_{i}^{occ.}\sum_{j}^{unocc.}\frac{1}{\Delta^{3}E_{i\to j}}$$

$$\times \langle \Psi_i | \delta(\mathbf{r}_N) | \Psi_j \rangle \langle \Psi_j | \delta(\mathbf{r}_{N'}) | \Psi_i \rangle. \tag{4.35}$$

If the LCAO approximation is used, equation (4.35) becomes

$$J^c_{NN'} = -\frac{8\mu_0^2\mu_B^2\hbar}{9\pi}\gamma_N\gamma_{N'}\Bigg|\sum_i^{\text{occ.}}\sum_j^{\text{unocc.}}\frac{1}{\Delta^3 E_{i\to j}}$$

$$\times \sum_\lambda\sum_\mu\sum_\nu\sum_\sigma {}^{\iota}C_{i\lambda}C_{j\mu}C_{j\nu}C_{i\sigma}\langle\chi_\lambda|\delta(\mathbf{r}_N)|\chi_\mu\rangle$$

$$\times \langle\chi_\nu|\delta(\mathbf{r}_{N'})|\chi_\sigma\rangle \tag{4.36}$$

Here, some approximations for the contact term are adopted such that χ_λ and χ_μ are s orbitals on atom N, χ_ν and χ_σ are s orbitals on atom N', and the most predominant terms are those involving the valence shell s orbital functions.

Using these approximations, we can rewrite equation (4.36) as

$$J^c_{NN'} = -\frac{8\mu_0^2\mu_B^2\hbar}{9\pi}\gamma_N\gamma_{N'}\langle\chi_{sN}|\delta(\mathbf{r}_N)|\chi_{sN}\rangle$$

$$\times \langle\chi_{sN'}|\delta(\mathbf{r}_{N'})|\chi_{sN'}\rangle \sum_i^{\text{occ.}}\sum_j^{\text{unocc.}}\frac{1}{\Delta^3 E_{i\to j}}$$

$$C_{isN}C_{jsN}C_{isN'}C_{jsN'}$$

$$= -\frac{8\mu_0^2\mu_B^2\hbar}{9\pi}\gamma_N\gamma_{N'}\langle\chi_{sN}|\delta(\mathbf{r}_N)|\chi_{sN}\rangle$$

$$\times \langle\chi_{sN'}|\delta(\mathbf{r}_{N'})|\chi_{sN'}\rangle \Pi_{NN'} \tag{4.37}$$

where

$$\Pi_{NN'} = \sum_i^{\text{occ.}}\sum_j^{\text{unocc.}}\frac{1}{\Delta^3 E_{i\to j}}C_{isN}C_{jsN}C_{isN'}C_{jsN'}. \tag{4.37a}$$

It is noted that $\Pi_{NN'}$, the mutual atom–atom polarizability, and $J^c_{NN'}$ can have either sign even though correlation between electrons of opposite spin is absent in this description.

If we assume an AEE value, ΔE, equation (4.37) becomes

$$J^c_{NN'} = -\frac{8\mu_0^2\mu_B^2\hbar}{9\pi}\gamma_N\gamma_{N'}\langle\chi_{sN}|\delta(\mathbf{r}_N)|\chi_{sN}\rangle$$

$$\times \langle\chi_{sN'}|\delta(\mathbf{r}_{N'})|\chi_{sN'}\rangle \frac{1}{\Delta E}\sum_i^{\text{occ.}}\sum_j^{\text{unocc.}} C_{isN}C_{jsN}C_{isN'}C_{jsN'}$$

$$= -\frac{8\mu_0^2\mu_B^2\hbar}{9\pi}\gamma_N\gamma_{N'}\frac{1}{\Delta E}\langle\chi_{sN}|\delta(\mathbf{r}_N)|\chi_{sN}\rangle$$

$$\times \langle\chi_{sN'}|\delta(\mathbf{r}_{N'})|\chi_{sN'}\rangle P^2_{s_Ns_{N'}} \tag{4.38}$$

where $P^2_{s_Ns_{N'}}$ is the s_N–$s_{N'}$ bond order.

In a similar way, the dipolar contribution can be developed as follows:

$$J^d_{NN'} = -\frac{\mu_0^2\mu_B^2 h}{24\pi^3}\gamma_N\gamma_{N'}\sum_i^{\text{occ.}}\sum_j^{\text{unocc.}}\frac{1}{\Delta^3 E_{i\to j}}$$

$$\times \sum_\lambda\sum_\mu\sum_\nu\sum_\sigma C_{i\lambda}C_{j\mu}C_{j\nu}C_{i\sigma}\langle\chi_\lambda|\frac{3r_{N\alpha}r_{N\beta}-r_N^2\delta_{\alpha\beta}}{r_N^5}|\chi_\mu\rangle$$

$$\times \langle\chi_\nu|\frac{3r_{N'\alpha}r_{N'\beta}-r_{N'}^2\delta_{\alpha\beta}}{r_{N'}^5}|\chi_\sigma\rangle, \qquad (4.39)$$

where a summation convention is used for the tensor suffixes α and β. According to Pople and Santry's approximation that only one-centre integrals are involved and only atoms with p valence orbitals are considered, there will be no contribution to $J^d_{NN'}$ if either N or N' is a hydrogen atom. All one-electron matrix elements are evaluated with this approximation. An example of the evaluation of a contributing one-electron matrix element is given by

$$\langle 2p_{xN}|\frac{3r_{xN}^2-r_N^2}{r_N^5}|2p_{xN}\rangle = \tfrac{4}{5}\langle r^{-3}\rangle_N, \qquad (4.40)$$

where $\langle r^{-3}\rangle_N$ is the mean value of r^{-3} for the 2p orbitals on atom N. By using this description and an AEE approximation, we have

$$J^d_{NN'} = -\frac{\mu_0^2\mu_B^2 h}{30\pi^3}\gamma_N\gamma_{N'}\langle r^{-3}\rangle_N\langle r^{-3}\rangle_{N'}\frac{1}{\Delta E}$$

$$\times \{2(P^2_{x_Nx_{N'}}+P^2_{y_Ny_{N'}}+P^2_{z_Nz_{N'}})+3(P_{x_Nx_{N'}}$$

$$\times P_{y_Ny_{N'}}+P_{y_Ny_{N'}}P_{z_Nz_{N'}}+P_{z_Nz_{N'}}P_{x_Nx_{N'}})$$

$$-(P^2_{x_Ny_{N'}}+P^2_{y_Nx_{N'}}+P^2_{y_Nz_{N'}}+P^2_{z_Ny_{N'}}+P^2_{z_Nx_{N'}}+P^2_{x_Nz_{N'}})$$

$$+3(P_{x_Ny_{N'}}P_{y_Nx_{N'}}+P_{y_Nz_{N'}}P_{z_Ny_{N'}}+P_{z_Nx_{N'}}P_{x_Nz_{N'}})\}. \qquad (4.41)$$

The orbital contribution can be separated into the two terms $J^o_{NN'_1}$ and $J^o_{NN'_2}$ which correspond to the two interactions described by equations (4.11) and (4.12), respectively. The first term, $J^o_{NN'_1}$, does not involve excited states and is expressed in the LCAO approximation by

$$J^o_{NN'_1} = \frac{\mu_0^2 e^2 h}{48\pi^3 m}\gamma_N\gamma_{N'}\sum_i^{\text{occ.}}\langle\Psi_i|\frac{\mathbf{r}_N\cdot\mathbf{r}_{N'}}{r_N^3 r_{N'}^3}|\Psi_i\rangle$$

$$= \frac{\mu_0^2 e^2 h}{48\pi^3 m}\gamma_N\gamma_{N'}\sum_i^{\text{occ.}}\sum_\lambda\sum_\mu C_{i\lambda}C_{i\mu}$$

$$\times \langle\chi_\lambda|\frac{\mathbf{r}_N\cdot\mathbf{r}_{N'}}{r_N^3 r_{N'}^3}|\chi_\mu\rangle. \qquad (4.42)$$

On the other hand, the second term, $J^o_{NN'_2}$, is expressed within the LCAO approximation as

$$J^o_{NN'_2} = -\frac{\mu_0^2 \mu_B^2 \hbar}{6\pi^3} \gamma_N \gamma_{N'} \sum_i^{\text{occ.}} \sum_j^{\text{unocc.}} \frac{1}{\Delta^1 E_{i \to j}}$$

$$\times \langle \Psi_i | \frac{\mathbf{m}_N}{r_N^3} | \Psi_j \rangle \langle \Psi_j | \frac{\mathbf{m}_{N'}}{r_{N'}^3} | \Psi_i \rangle$$

$$= -\frac{\mu_0^2 \mu_B^2 \hbar}{6\pi^3} \gamma_N \gamma_{N'} \sum_i^{\text{occ.}} \sum_j^{\text{unocc.}} \frac{1}{\Delta^1 E_{i \to j}}$$

$$\times \sum_\lambda \sum_\mu \sum_\nu \sum_\sigma C_{i\lambda} C_{j\mu} C_{j\nu} C_{i\sigma} \langle \chi_\lambda | \frac{\mathbf{m}_N^o}{r_N^3} | \chi_\mu \rangle$$

$$\times \langle \chi_\nu | \frac{\mathbf{m}_{N'}^o}{r_N^3} | \chi_\sigma \rangle. \tag{4.43}$$

This term involves the mixing of the ground state spin singlet function and excited state singlet functions, $^1\Psi_{i \to j}$. By employing the approximations used for the calculation of the dipolar term, we have

$$\langle 2p_{xN} | \frac{\mathbf{m}_{Nz}^o}{r_N^3} | 2p_{yN} \rangle = -i \langle r^{-3} \rangle_N. \tag{4.44}$$

If we use relations such as equation (4.44) and an AEE approximation, equation (4.43) becomes

$$J^o_{NN'_2} = -\frac{\mu_0^2 \mu_B^2 \hbar}{6\pi^3} \gamma_N \gamma_{N'} \langle r^{-3} \rangle_N \langle r^{-3} \rangle_{N'} \frac{1}{\Delta E}$$

$$\times \{ P_{x_N x_N'} P_{y_N y_{N'}} + P_{y_N y_{N'}} P_{z_N z_{N'}} + P_{z_N z_{N'}} P_{x_N x_{N'}}$$

$$- P_{x_N y_{N'}} P_{y_N x_{N'}} - P_{y_N z_{N'}} P_{z_N y_{N'}} - P_{z_N x_{N'}} P_{x_N z_{N'}} \}. \tag{4.45}$$

At this level of approximation, the orbital term is zero if either of the coupled atoms is hydrogen.

Within the framework of the SCF MO theory, the triplet and singlet excitation energies are given in terms of the simple differences of the orbital energies and J_{ik} and K_{jk}. Thus the excitation energies are expressed by

$$\Delta^1 E_{k \to j} (= E_k - E_j) = \varepsilon_k - \varepsilon_j - J_{jk} + 2K_{jk}, \tag{4.46}$$

$$\Delta^3 E_{k \to j} (= E_k - E_j) = \varepsilon_k - \varepsilon_j - J_{jk}, \tag{4.47}$$

where ε is the appropriate orbital energy.

4.B.2 FPT Method

The FPT method described in Section 2.A.3 may be applied to calculate nuclear spin–spin couplings.[13] In this method, a perturbation corresponding to one of the three coupling mechanisms is introduced during the calculation of the SCF wavefunction for a given molecule.

The Hamiltonian involving the *contact* interaction is given by

$$\mathcal{H} = \mathcal{H}_0 + \frac{4\mu_0\mu_B\hbar}{3}\sum_k\sum_i \delta(\mathbf{r}_{kN})\mathbf{S}_k\cdot\boldsymbol{\mu}_i, \qquad (4.48)$$

where the first term represents the unperturbed system. This equation can be rewritten using two nuclear moments μ_N and $\mu_{N'}$ both directed along the z axis, which interact with each other:

$$\mathcal{H} = \mathcal{H}_0 + \mu_N\mathcal{H}_N + \mu_{N'}\mathcal{H}_{N'}, \qquad (4.49)$$

where

$$\mathcal{H}_N = -\frac{4\mu_0\mu_B\hbar}{3}\sum_k \delta(\mathbf{r}_{kN})\mathbf{S}_{k_z}, \qquad (4.50)$$

$$\mathcal{H}_{N'} = -\frac{4\mu_0\mu_B\hbar}{3}\sum_k \delta(\mathbf{r}_{kN'})\mathbf{S}_{k_z}. \qquad (4.51)$$

Using equations (4.49)–(4.51) and equations (2.35), (2.41) and (2.146) the resulting expression for the contact contribution to the spin–spin coupling is given by

$$J^c_{NN'} = \frac{4\mu_0^2\mu_B^2\hbar}{9\pi}\gamma_N\gamma_{N'}\left[\frac{\partial}{\partial\mu}\langle\Psi(\mu_N)|\mathcal{H}_{N'}|\Psi(\mu_{N'})\rangle\right]_{\mu_{N'}=0}, \qquad (4.52)$$

where $\Psi(\mu_{N'})$ is the wavefunction for the molecule in the presence of a perturbing nuclear spin of atom N' with nuclear moment $\mu_{N'}$. We will develop a method of calculating spin–spin couplings based on equation (4.52).

For convenience, we define a spin-density matrix in terms of the following difference:

$$\mathbf{P} = \mathbf{P}^\alpha - \mathbf{P}^\beta. \qquad (4.53)$$

The elements of the matrices, $\mathbf{P}^\alpha$ and $\mathbf{P}^\beta$, are defined in Section 2.B.4. The SCF equations involving the contact interaction are given by modifying equations (2.66)–(2.69) as follows:

$$F^\alpha_{\mu\nu} = \mathcal{H}^{core}_{\mu\nu} + \frac{2\mu_0\mu_B\hbar}{3}\int \chi^*_\mu \delta(\mathbf{r}_N)\chi_\nu\, dv$$

$$+ \sum_\lambda\sum_\sigma \{P_{\lambda\sigma}\langle\mu\nu|\lambda\sigma\rangle - P^\alpha_{\lambda\sigma}\langle\mu\lambda|\nu\sigma\rangle\}, \qquad (4.54)$$

$$F^{\beta}_{\mu\nu} = \mathcal{H}^{core}_{\mu\nu} - \frac{2\mu_0 \mu_B \hbar}{3} \int \chi_\mu \delta(\mathbf{r}_N) \chi_\nu \, dv$$

$$+ \sum_\lambda \sum_\sigma \{P_{\lambda\sigma}\langle\mu\nu|\lambda\sigma\rangle - P^\beta_{\lambda\sigma}\langle\mu\lambda|\nu\sigma\rangle\}. \tag{4.55}$$

Through these SCF equations we can obtain an expression for the contact contribution to the spin–spin coupling:

$$J^c_{NN'} = \frac{4\mu_0^2 \mu_B^2 \hbar}{9\pi} \sum_\mu \sum_\nu \int \chi_\mu \delta(\mathbf{r}_N) \chi_\nu \, dv$$

$$\times \left[\frac{\partial}{\partial \mu_{N'}} P_{\mu\nu}(\mu_{N'})\right]_{\mu_{N'}=0} \tag{4.56}$$

Practically, this equation can be evaluated using unrestricted LCAO-SCF wavefunctions. The integral in equation (4.56) is replaced within the framework of the INDO approximation by

$$\int \chi^*_\mu \delta(\mathbf{r}_{N'}) \chi_\nu \, dv = S_{N'}(0)^2, \tag{4.57}$$

where $S_{N'}(0)^2$ is the density at the nucleus N' of the valence s electrons, and if either χ^*_μ or χ_ν are not s valence orbitals on atom N', then the integral vanishes. The s-orbital densities for various free atoms are given in Table 4.1, as estimated by Pople et al.[13]

Table 4.1 s-orbital densities† at the nucleus for free atoms calculated by Pople et al.[12]

Nucleus	$S_N(0)^2$
H	0.3724
B	2.2825
C	4.0318
N	6.9265
O	12.0658
F	21.3126

† Values are given in terms of $(a_0)^{-3}$, where a_0 is the Bohr radius.

Therefore, according to the INDO approximation, equation (4.56) becomes

$$J^c_{NN'} = \frac{4\mu_0^2\mu_B^2\hbar}{9\pi} \gamma_N\gamma_{N'} S_N(0)^2 S_{N'}(0)^2$$
$$\times \left[\frac{\partial}{\partial h_{N'}} P_{S_N S_{N'}}(h_{N'})\right]_{h_{N'}=0}, \qquad (4.58)$$

where

$$h_{N'} = \frac{4\mu_0\mu_B}{3} \hbar\mu_{N'} S_{N'}(0)^2. \qquad (4.59)$$

This method is a powerful and practical approach to the calculation of spin–spin couplings and, in combination with the simplified integral treatment used in the INDO approximation, it becomes possible to perform calculations on large molecules within a reasonable amount of computer time.

Furthermore, we can say that according to an independent electron MO approximation, the derivative in equation (4.58) is equivalent to the mutual polarizability, $\Pi_{S_N S_{N'}}$, being defined by equation (4.37a).

4.B.3 SCP Method

In the previous section, the FPT method for estimating the contact term has been described. This method can be relatively time-consuming for large molecules. Blizzard and Santry[14] have presented a more efficient type of procedure for calculating all three coupling mechanisms. This is the self-consistent perturbation method (SCP), which is now described briefly.

If a LCAO wavefunction is used, the derivatives in equation (4.58) can be calculated from the first-order perturbation, $C^{(1)}$, to the MO coefficient matrix, C, and $C^{(1)}$ is obtained from the unperturbed coefficient matrix, $C^{(0)}$, as

$$C^{(1)}_j = -\left[\sum_l^{\text{unocc.}} (\varepsilon_l^{(0)} - \varepsilon_j^{(0)})^{-1} C_l^{(0)*} C_l^{(0)}\right] F^{(1)} C_j^{(0)}, \qquad (4.60)$$

where $F^{(1)}$ is the first-order change in the Fock matrix due to the perturbation under consideration.

In the case of the *contact* term, the expression for the second-order energy given by equation (2.146) is written through the unrestricted SCF procedure by

$$E_{NN'} = \sum_\sigma \sum_v H^{N'(1)}_{\sigma v} \left(\frac{\partial P_{\sigma v}}{\partial \mu_N}\right)_{\mu_N=0}, \qquad (4.61)$$

where

$$H^{N'(1)}_{\sigma v} = \langle \chi_\sigma | \mathscr{H}_{N'} | \chi_v \rangle. \qquad (4.62)$$

Thus, the first-order change in $\mathbf{P}^\alpha$, with respect to μ_N, is given by

$$P_{\sigma\mu}^{\alpha(1)} = \left(\frac{\partial P_{\sigma\nu}^\alpha}{\partial \mu_N}\right)_{\mu_N=0}$$

$$= \sum_{i=1}^{n} (C_{\sigma i}^{\alpha(1)*}C_{vi}^{(0)} + C_{\sigma i}^{(0)}C_{vi}^{(1)}), \qquad (4.63)$$

where the superscripts (1) and (0) refer to perturbed and unperturbed quantities, respectively. Therefore, the contact contribution to the second-order energy is given, in the INDO approximation, by

$$E_{NN'} = -\frac{16\mu_0^2\mu_B^2\hbar^2}{9} S_N(0)^2 S_{N'}(0)^2 P_{S_N S_{N'}}^{\alpha(1)}. \qquad (4.64)$$

Using this equation, the expression for the contact term becomes

$$J_{NN'}^c = -\frac{8\mu_0^2\mu_B\hbar}{9\pi} \gamma_N\gamma_{N'} S_N(0)^2 S_{N'}(0)^2 P_{S_N S_{N'}}^{\alpha(1)}. \qquad (4.65)$$

In the case of the orbital term, since the corresponding interaction is not isotropic, the calculation must be carried out with the perturbing nuclear spin, μ_N, in each of the three Cartesian directions. Therefore, the isotropic *orbital* contribution to $J_{NN'}$ is obtained from the average of the results of three calculations. The expression for the orbital term is given by

$$J_{NN'}^o = \frac{\mu_0^2 e^2 \hbar}{24\pi^3 m} - \gamma_N\gamma_{N'} \langle r^{-3}\rangle_N \langle r^{-3}\rangle_{N'} (P_{x_N y_N} + P_{x_N z_N} + P_{y_N z_N}). \qquad (4.66)$$

In the case of the *dipolar* interaction, the treatment is the most complicated when compared with those for the other coupling terms because it causes electron spin polarization as well as both real and imaginary perturbations to the wavefunction. The dipolar contribution is expressed as the sum of three terms such as the second-order energy contributions to the $\alpha\alpha$ and $\beta\beta$ spin components, the real $\beta\alpha$ and $\alpha\beta$ components, and the imaginary $\beta\alpha$ and $\alpha\beta$ components of the dipolar term. Thus, the expression for the dipolar contribution is given by

$$J_{NN'}^d = -\frac{\mu_0^2\mu_B^2\hbar}{30\pi^3} \gamma_N\gamma_{N'} \langle r^{-3}\rangle_N \langle r^{-3}\rangle_{N'}$$

$$\times (2P_{z_N z_{N'}}^{\alpha\alpha(1)} - P_{x_N' x_N}^{\alpha\alpha(1)} - P_{y_N' y_N}^{\alpha\alpha(1)}$$

$$+ 3P_{x_N' z_N}^{\alpha\beta(1)} + 3Q_{y_N' z_N}^{\alpha\beta(1)}) \qquad (4.67)$$

where $Q^{\alpha\beta(1)}$ is the imaginary part of the bond order matrix and is given by

$$Q_{v\sigma}^{\alpha\beta(1)} = i\sum_j^\alpha C_{vj}^{\alpha\alpha(0)}C_{\sigma j}^{\beta\alpha(1)} - i\sum_j^\beta C_{vj}^{\alpha\beta(1)}C_{\sigma j}^{\beta\beta(0)}. \qquad (4.68)$$

4.C VB APPROACHES

A general VB approach to the calculation of spin–spin couplings has been developed by Karplus and coworkers.[15,16] This approach has had considerable success in the interpretation of the dependence of *vicinal* couplings on dihedral angle and of *geminal* couplings on the bond angle between the coupled nuclei. Nevertheless, mathematical difficulties have made the VB method, in general, less readily applicable than the MO procedure for calculating spin–spin couplings.

Until recently, all VB approaches had been based on the AEE approximation. Therefore, a general VB expression for spin–spin coupling is developed within the confines of the AEE approximation.

If we consider a system of $2n$ electrons distributed among $2n$ singly occupied orbitals $\varphi_1, \varphi_2, \ldots, \varphi_{2n}$, then the singlet state function, $^1\Psi$, as a linear combination of independent valence bond structures, can be expressed as implied in equation (2.118) by

$$^1\Psi = \sum_i C_i \Psi_i, \qquad (4.69)$$

where

$$\Psi_i = \frac{1}{\sqrt{2n}} \sum_R (-1)^R R \left[\frac{1}{\sqrt{2n}} \sum_P (-1)^P P \psi_1(1)\alpha(1) \right.$$

$$\left. \times \varphi_2(2)\beta(2), \ldots, \varphi_{2n}(2n)\beta(2n) \right). \qquad (4.70)$$

For canonical structures which have a spin function α associated with φ_1, and so on, a construction of the complete set of singlet functions and the evaluation of the coefficients of the Coulomb and exchange integrals can be carried out by the Rumer–Pauling schemes.[17,18]

Following Ramsey,[9,10] the *contact* term is expressed by equation (4.8) with an AEE approximation. A discussion of the procedure for the estimation of the matrix element $\langle 0| \delta(\mathbf{r}_{kN}) \delta(\mathbf{r}_{kN'}) \mathbf{S}_k \mathbf{S}_j |0\rangle$, by using equation (4.70), is the main aim in this section. Rearrangement of equation (4.26) gives

$$P_{kj} = \tfrac{1}{2}(1 + 4\mathbf{S}_k\mathbf{S}_j). \qquad (4.71)$$

Substituting equations (4.69) and (4.71) into equation (4.8), we obtain

$$J^c_{NN'} = -\frac{\mu_0^2 \mu_B^2 \hbar}{9\pi} \gamma_N \gamma_{N'} \frac{1}{\Delta E}$$

$$\times \sum_i \sum_l C_i C_l \langle \Psi_i | \sum_k \sum_j \delta(\mathbf{r}_{kN}) \delta(\mathbf{r}_{kN'})$$

$$\times 2(P_{kj} - 1)|\Psi_1\rangle. \qquad (4.72)$$

4 SPIN–SPIN COUPLINGS

In equation (4.72), two types of integral remain to be evaluated,

$$I_1 = \langle \Psi_i | \sum_k \sum_j \delta(\mathbf{r}_{kN}) \delta(\mathbf{r}_{kN'}) | \Psi_l \rangle \tag{4.73}$$

and

$$I_2 = \langle \Psi_i | \sum_k \sum_j \delta(\mathbf{r}_{kN}) \delta(\mathbf{r}_{kN'}) P_{kj} | \Psi_l \rangle. \tag{4.74}$$

If equation (4.70) is substituted into equations (4.73) and (4.74), I_1 and I_2 reduce to the following expressions:

$$I_1 = \left\langle \left(\frac{1}{2^n}\right) \sum_R \sum_{R'} (-1)^{R+R'} \{ |\varphi_1(1)\alpha(1)\varphi_2(2)\beta(2) \cdots \right.$$
$$\times \varphi_{2n}(1n)\beta(2n)|\} |\delta(\mathbf{r}_{kN}) \delta(\mathbf{r}_{jN'})|$$
$$\left. \times \{|\varphi_1(1)\alpha(1)\varphi_2(2)\beta(2) \cdots \varphi_{2n}(2n)\beta(2n)|\} \right\rangle \tag{4.75}$$

and

$$I_2 = \left\langle \left(\frac{1}{2^n}\right) \sum_R \sum_{R'} (-1)^{R+R'} \{ |\varphi_1(1)\alpha(1)\varphi_2(2)\beta(2) \cdots \right.$$
$$\times \varphi_{2n}(2n)\beta(2n)|\} |\delta(\mathbf{r}_{kN}) \delta(\mathbf{r}_{jN'}) P_{kj}|$$
$$\left. \times \{|\varphi(1)\alpha(1)\varphi_2(2)\beta(2) \cdots \varphi_{2n}(2n)\beta(2n)|\} \right\rangle. \tag{4.76}$$

In equation (4.75) I_1 is in the form of a valence bond Coulomb integral with $\delta(\mathbf{r}_{kN}) \delta(\mathbf{r}_{jN'})$ replacing the Hamiltonian operator. Thus, we have

$$I_1 = \left\langle \left(\frac{1}{2^{n-I_{il}}}\right) \{|\varphi_1(1)\alpha(1)\varphi_2(2)\beta(2) \cdots \varphi_{2n}(2n)\beta(2n)|\} \right.$$
$$\left. \times |\delta(\mathbf{r}_{kN}) \delta(\mathbf{r}_{jN'})| \{|\varphi_1(1)\alpha(1)\varphi_2(2)\beta(2) \cdots \varphi_{2n}(2n)\beta(2n)|\} \right\rangle$$
$$= \left(\frac{1}{2^{n-I_{il}}}\right) S_N(0)^2 S_{N'}(0)^2, \tag{4.77}$$

where I_{il} is the number of islands in the superposition scheme for structures i and l, and $S_N(0)^2$ and $S_{N'}(0)^2$ are the electron densities at nuclei N and N', respectively. On the other hand, the spin exchange operator, P_{kj}, in I_2 (equation (4.76)) has the same effect on the spin orthogonality of the integral. Thus, we can write

$$I_2 = -\left[\frac{f_{il}(P_{NN'})}{2^{n-I_{il}}}\right] S_N(0)^2 S_{N'}(0)^2, \tag{4.78}$$

where $f_{il}(P_{NN'})$ is the exchange factor for the nuclei N and N' in the superposition scheme for the structures i and l. Substituting equations (4.77) and (4.78) into (4.72), we have

$$J^c_{NN'} = -\frac{8\mu_0^2\mu_B^2\hbar}{27\pi}\gamma_N\gamma_{N'}\frac{1}{\Delta E}S_N(0)^2 S_{N'}(0)^2$$

$$\times \sum_i \sum_l C_i C_l \left(\frac{1}{2^{n-l_{il}}}\right)[1 + 2f_{il}(P_{NN'})]. \quad (4.79)$$

In the case of non-canonical structures, equation (4.79) is not applicable because of a different method of estimation of f_{il} in equation (4.79), thus instead of equation (4.79) we must use

$$J^c_{NN'} = -\frac{8\mu_0^2\mu_B^2\hbar}{27\pi}\gamma_N\gamma_{N'}\frac{1}{\Delta E}S_N(0)^2 S_{N'}(0)^2$$

$$\times \sum_i \sum_l C_i C_l \left[\frac{2(-1)^{r'_{il}}}{2^{n-l'_{il}}} - \frac{(-1)^{r_{il}}}{2^{n-l_{il}}}\right], \quad (4.80)$$

where the primed and unprimed indices refer to the scheme from the permuted and unpermuted structures, respectively. In equation (4.80) r_{il} and r'_{il} refer to the number of spin reversals required for an initial satisfactory orientation in the superposition scheme.

A VB theory of spin–spin couplings, without the AEE approximation, has been formulated by Barfield.[19,20] Using the transition spin-density matrix formulation, an expression for the contact contribution to the coupling is given by

$$J^c_{NN'} = -\frac{4\mu_0^2\mu_B^2\hbar}{9\pi}\gamma_N\gamma_{N'}\sum_{k(\neq 0)}\frac{Q(0\,k_0|1_N, 1_N)Q(0\,k_0|1_{N'}, 1_{N'})}{E_k - E_0} \quad (4.81)$$

where $Q(0\,k_0|1_N, 1_N)$ is the transition spin-density matrix, evaluated at N, between the ground state and the triplet state k, with energy E_k, and is expressed by

$$Q(0\,k_0|1_N, 1_N) = \sum_j \sum_l C_j C_{kl} \frac{1}{2^{n-l_{il}}} \sum_h r(1)r(1)^* f^h_{jl}. \quad (4.82)$$

The notations in this equation are similar to those in equation (4.79); C_{kl} is the coefficient of the structure l in the triplet k, and h refers to the electrons concerned.

4.D CALCULATIONS OF SPIN–SPIN COUPLINGS

In this section emphasis is placed on a comparison of calculated and experimental spin–spin coupling data for various nuclei, N and N', separated by n intervening bonds. The resulting coupling of which is denoted by $^nJ(N-N')$.

4.D.1 SOS Calculations

We begin with some examples of *ab initio* calculations of spin–spin coupling. For the hydrogen molecule a ground state wavefunction of good accuracy is available. Therefore, a lot of *ab initio* approaches to the spin–spin coupling interaction in the hydrogen molecule, H_2, have been performed. The SOS calculation may be carried out by using an extended basis set including

Table 4.2 Values of spin–spin couplings, in hertz, obtained from SOS calculations within the INDO framework

	^nJ		
	Without CI	With CI	Experiment
$^1H-^1H$ coupling			
^2J			
Methane	−2.72	−1.49	−12.4
Ethylene	8.81	9.79	2.3
Formaldehyde	36.55	43.99	40.2
^3J			
Acetylene	10.97	12.29	9.8
Ethylene (*cis*)	9.33	11.99	11.5
Ethylene (*trans*)	25.38	32.26	19.1
Ethane (*trans*)	18.88	24.46	
Ethane (*gauche*)	2.75	3.66	
Ethane (average)	8.13	10.59	8.0
$^{13}C-^1H$ coupling			
^1J			
Methane	70.54	92.7	125.0
Ethylene	86.8	119.8	156.0
Formaldehyde	112.9	139.7	172.0
^2J			
Acetylene	7.57	7.26	49.3
Ethane	0.42	−1.66	−4.5

configuration interaction, CI, between the ground and doubly-excited states as well as CI with singly-excited triplet states. The result obtained is 312 Hz for $^1J(^1H-^1H)$.[21] An SOS calculation employing MOs obtained from a basis set of 32 GTOs gives 200 Hz for the coupling.[22] The agreement with the experimental value of 278 Hz[23] (this value is obtained from the experimental value for HD after multiplying by $\gamma_H/\gamma_D = 6.514$) is reasonable.

For the $^{19}F-^1H$ coupling of the HF molecule the *ab initio* calculation, using a basis set of 14 STOs, leads to a value of 621.3 Hz for $^1J(^{10}F-^1H)$.[24] This agrees well with the experimental value (530 Hz).[25] These works have at least succeeded in predicting the correct sign and magnitude of the coupling interaction. However, *ab initio* calculations are only possible for small molecules at present. For large molecules, the majority of calculations reported to date are semi-empirical ones, involving the SOS, FPT and SCP methods. Some examples which provide a satisfactory account of spin–spin couplings by the SOS approach, within the INDO framework, are presented in Table 4.2.[26] In the case of $^1H-^1H$ couplings, calculated $2J(^1H-^1H)$ values compare badly with experiment, but calculations are reasonably successful in the case of $^3J(^1H-^1H)$ couplings. The INDO-CI procedure appears to give the most satisfactory results. For $^{13}C-^1H$ couplings, the calculated results are not good, but the calculation is able to predict both the sign of the coupling and the experimentally observed trend in values.

4.D.2 FPT Calculations

FPT calculations for spin–spin couplings have been widely carried out in both the INDO and CNDO/2 approximations. For the purpose of presenting a general discussion of calculated and experimental results the s-orbital densities are often treated as adjustable parameters in order to obtain the best overall fit of the calculated couplings to the available experimental data. The values of the s-orbital densities on free atoms proposed by Pople *et al.*[13] are shown in Table 4.1. Some examples of the calculated values of spin–spin couplings between ^1H and ^1H, ^1H and ^{13}C, and between ^1H and ^{19}F are given together with the corresponding experimental values in Table 4.3. It is clear that calculations within the INDO framework provide better agreement with the experimental data than do those obtained within the CNDO/2 framework.

Calculated values for $^1J(^{13}C-^1H)$ are positive and increase in magnitude along the series ethane, ethylene and acetylene as expected from a consideration of the approximate proportionality of the couplings to the s character of the C—H bond, obtained from a simple hybridization picture. The values for *geminal* H–H couplings are negative in tetrahedral systems and are positive in ethylene and similar cases containing trigonally hybridized

Table 4.3 Values of spin–spin couplings, in hertz, obtained from FPT calculations within the INDO and CNDO/2 frameworks

		^nJ		
		INDO	CNDO/2	Experiment
^1J				
	Methane H–C	122.92	93.19	125.0
	Ethane H–C	122.12	93.30	124.9
	Ethylene H–C	156.71	127.63	156.4
	Acetylene H–C	205.49	232.65	248.7
	Benzene H–C	140.29	116.00	157.5
^2J				
	Water H–O–H	−8.07	1.31	−7.2
	Ammonia H–N–H	−6.37	1.60	−10.4
	Methane H–C–H	−6.13	1.17	−12.4
	Cyclopropane H–C–H	−0.40		−4.3
	Acetonitrile H–C–H	−7.73		−16.9
	Ethylene H–C–H	3.25	8.48	2.5
	Benzene H–C–H	−4.94	−0.17	1.0
^3J				
	Ethane H–C–C–H (trans)	18.63	15.43	
	Ethane H–C–C–H (gauche)	3.25	2.43	
	Ethane (average)	8.38	6.76	8.0
	Ethylene H–C–C–H (cis)	9.31	8.04	11.7
	Ethylene H–C–C–H (trans)	25.15	19.50	19.1
	Benzene C–C–C–H	9.40	5.51	7.4
^4J				
	Benzene H–C–C–H	2.13	1.90	1.37
	Benzene H–C–C–C–C	−2.27	−0.06	−1.1
	Allene H–C–C–C–H	−9.69		−7.0

carbons. The values of *vicinal* ^{1}H–^{1}H couplings are positive and those for the *trans* form are greater than are those in the *gauche* or *cis* forms. The values for *vicinal* ^{13}C–^{1}H couplings are positive. In the case of long-range ^{1}H–^{1}H couplings, the value for allene is negative and large, but $^4J(^1\text{H}-^1\text{H})$ in benzene is positive. The $^4J(^{13}\text{C}-^1\text{H})$ value in benzene is negative. Most experimental trends are well reproduced by the results of the INDO calculations.

Table 4.4 Values, in hertz, of the various components of the anisotropic part of the coupling between carbon and fluorine nuclei in methyl fluoride obtained by a FPT calculation

	Contributions to the coupling anisotropy				
	Contact–dipolar	Dipolar	Orbital	Total	Experiment
$^{13}C-^{1}H$	−19.0	0.0	0.0	−19.0	1890 ± 130
$^{19}F-^{13}C$	208.0	26.0	27.0	261.0	700 ± 130

NMR measurements employing nematic solvents often permit the estimation of spin–spin coupling anisotropy. The results of FPT-INDO calculations are compared with experimental ones for one-bond couplings in methyl fluoride in Table 4.4.[27] It is shown that the major contribution to the $^{19}F-^{13}C$ coupling anisotropy arises from the interaction of the dipolar and contact terms. Although the calculated values of the isotropic couplings are in reasonable agreement with the experimental results, the anisotropies agree less well.

4.D.3 SCP Calculations

For couplings involving protons it is sufficient to consider only the contact contribution, as has been demonstrated by means of FPT calculations. It might be worthwhile to mention that this is due to the s electron dependence. However, for couplings not involving protons, such as $^{13}C-^{13}C$, $^{15}N-^{13}C$, $^{19}F-^{13}C$, $^{19}F-^{19}F$, etc., the dipolar and orbital contributions must also be considered. The SCP calculated values of the various components for some $^{13}C-^{13}C$, $^{15}N-^{13}C$, $^{19}F-^{13}C$ and $^{19}F-^{19}F$ couplings by means of INDO parameter for some molecules are shown in Table 4.5.[14,28] For $^{13}C-^{13}C$ couplings, the calculated results indicate that a reasonably good description may be obtained by considering the contact term alone. However, the inclusion of the dipolar and orbital terms provides an improvement in the agreement between calculation and experiment. The orbital term is of opposite sign to the contact term in all cases considered, except that of acetylene. For $^{15}N-^{13}C$ coupling, the contact, dipolar and orbital terms can each provide the dominant contribution to the observed couplings. The sign of the $^{15}N-^{13}C$ coupling in CH_3NH_2 and in acetonitrile is negative. In CH_3NC the sign of the $^1J(^{15}N-^{13}C)$ coupling has been determined to be negative experimentally, which is in agreement with the, non-contact-dominated, calculated result.[24] In the calculation of $^{19}F-^{13}C$ and $^{19}F-^{19}F$ couplings, the inclusion of the dipolar and orbital terms greatly improves the agreement between calculation and experiment. The contact and orbital terms are of the same sign as the experimentally determined couplings.

Table 4.5 Values of the various components of $^{13}C-^{13}C$, $^{15}N-^{13}C$ and $^{19}F-^{13}C$ couplings, in hertz, for some molecules obtained by SCP INDO calculations

	Coupled nuclei	J^c	J^d	J^o	Total	Experiment
^1J						
Acetylene	$^{13}C-^{13}C$	140.80	8.31	23.59	172.70	171.5
Ethylene	$^{13}C-^{13}C$	70.61	3.92	−18.58	55.95	67.6
Acetonitrile	$^{13}C-^{13}C$	65.96	0.57	−2.56	63.97	65.96
Ethane	$^{13}C-^{13}C$	35.63	0.73	−2.91	33.45	55.63
CH_3NH_2	$^{15}N-^{13}C$	−2.33	−0.20	0.41	−2.12	−4.5
Acetonitrile	$^{15}N-^{13}C$	2.33	−14.95	−9.29	−21.91	−17.5
CH_3NC	$^{15}N-^{13}C$	14.85	−7.89	−12.44	−5.48	−8.8
CH_3F	$^{19}F-^{13}C$	−168.66	30.99	−65.00	−202.68	−157.5
CH_2F_2	$^{19}F-^{13}C$	−164.45	23.01	−118.20	−259.65	−234.8
CHF_3	$^{19}F-^{13}C$	−152.56	17.09	−149.70	−285.16	−274.3
CHF_3	$^{19}F-^{19}F$	36.29	20.11	47.28	103.68	150.0
$gem\text{-}C_2HF_2$	$^{19}F-^{19}F$	5.43	19.94	56.37	81.74	87.0
$trans\text{-}C_2HF_3$	$^{19}F-^{19}F$	−42.25	5.19	−64.91	−101.97	−119.0

4.D.4 Comparison of the Results from Different Theories and MO Treatments

We now consider a comparison of the calculations of spin couplings by means of different theories and MOs. The calculated values are, sometimes, dependent on the chosen theory and MO treatment used. Thus, it is important to know the features of the different theories and MOs used in the calculation of spin–spin couplings. Some examples of the calculations on spin–spin couplings by means of different theories and MOs are listed in Table 4.6.

In non-empirical calculations, the coupled Hartree–Fock method overestimates the values of spin–spin couplings, but the FPT-CI method reproduces the experimental value well. Turning to the area of semi-empirical calculations, the calculations, in general, reproduce the experimental values well except for those employing the FPT-MINDO/3 and SOS-CNDO procedures. For a series of substituted methanes, the FPT-INDO, SOS-INDO and SCP-INDO methods, as well as the non-empirical calculations, are capable of reproducing the effects on the couplings of electronegative substituents. In addition, the signs of spin–spin couplings are correctly predicted except for the case of the coupled Hartree–Fock calculation of CH_3CN. On the whole, it is shown that the non-empirical FPT-CI method and theories using the semi-empirical INDO method reproduce the experimental values well.

Some non-empirical calculations have recently appeared for molecules of medium size. The gap between non-empirical and semi-empirical calculations is closing in terms of the size of molecule which we can handle. Further developments of non-empirical calculations will probably provide deeper insight into our understanding of spin–spin couplings.

4.D.5 Correlations of Spin–Spin Couplings with Other Molecular Properties

Spin–spin couplings are usually dependent on various molecular parameters such as hybridization, electronegativity of substituents and geometries. Thus we can obtain some information on these properties through the interpretation of spin–spin couplings.

(a) *Hybridization Effects*

These effects have been widely considered since the nature of bonding can be understood by means of spin–spin coupling, when the contact term is of greatest importance. It has been suggested that the spin–spin coupling

Table 4.6 Calculated values, in hertz, of spin couplings in some molecules by means of different theories and MOs compared with experimental results

	Non-empirical		FPT		Semi-empirical			Experimental
	Coupled Hartree–Fock[30]	FPT-CI[30]	INDO[13,31,32]	MINDO/3[33]	SCP INDO[29]	SOS INDO[35]	CNDO[36]	
$^1J(^{13}C-^1H)$								
CH_4	150.7	120.1	122.9	140.3	120.9	120.9		125.0
CH_3CH_3	157.7	121.0	122.1	99.2		124.3		124.9
CH_3OH	169.5		135.3			152.4		141.0
Benzene			140.3	134.1			57.13	157.5
CH_3F	177.8	136.6	140.1	122.3		171.4		149.1
CH_2CH_2			156.7	132.9		171.8		156.4
$^1J(^{15}N-^{13}C)$								
CH_3NH_2	−11.6	−3.3			−2.1			−4.5
CH_3CN	4.4[34]				−16.8			−17.5

depends upon the amount of s-character in the bond joining the nuclei. An empirical relationship (4.83) has been proposed,[37] i.e.

$$k|^1J(^{15}N-^{13}C)| = \%S_C \%S_N, \qquad (4.83)$$

between the coupling and percentage s-character of the atomic orbitals forming the σ bond between the ^{15}N and ^{13}C nuclei. FPT-INDO calculations show that the proportionality constant, k, in equation (4.83) should be 94 and the sign of the coupling should be negative.

For $^{13}C-^{13}C$ couplings, Newton and coworkers[38-40] showed that the experimental results for a number of highly strained ring systems fit the linear relationship

$$^1J(^{13}C_1-^{13}C_2) = 0.0621 \%S_{C_1} \%S_{C_2} - 10.2. \qquad (4.84)$$

For $^{13}C-^1H$ couplings, Müller and Pritchard[41] have suggested the following relationship:

$$^1J(^{13}C-^1H) = 500 \%S_{CH}. \qquad (4.85)$$

This simple relationship has been modified by some later workers.

In general, X–H couplings show a reasonable correlation with the amount of s-character in the $^{13}C-^1H$ bond. Substitution may cause large changes in the effective nuclear charge, in which case exact correlations are not anticipated. The relationship between $^1J(^{13}C-^1H)$ and the $^{13}C-^1H$ bond order is relatively unsuccessful in accounting for substituent effects, whereas comparison of the observed coupling with the calculated contact term is more satisfactory. FPT-INDO calculations for $^1J(^{13}C-^1H)$ have been carried out for various compounds. Good agreement is found for molecules with I^+ substituents such as F, OR, NR$_2$ and =O, but the situation is less satisfactory for molecules with I^- substituents such as CF$_3$, C(O)X, NO$_2$ and CN. It is possible that through-space effects may be important for some of these substituents.

(b) *Angular Dependence*

It is well known that some couplings show an angular dependence, especially two-bond and three-bond couplings which depend sensitively on the appropriate bond angles and dihedral angles, respectively.

For *geminal* $^1H-^1H$ couplings, experiments show that in hydrocarbons $^2J(^1H-^1H)$ increases with increasing H—C—H bond angle. An FPT-INDO calculation shows that an increase in the bond angle leads to an algebraic increase in the coupling in methane and ethylene, as shown in Table 4.7.[42] It is shown that *geminal* couplings are very sensitive to the choice of bond angle.

Since Karplus' prediction,[43] from VB theory, of a strong conformational

Table 4.7 Values of ^2J (^{1}H–^{1}H), in hertz, for methane and ethylene as a function of the H—C—H angle obtained by a FPT-INDO calculation

Angle	^2J (^{1}H–^{1}H) methane coupling	Angle	^2J (^{1}H–^{1}H) ethylene coupling
103°	−7.81	114°	1.70
107°	−6.69	118°	2.75
111°	−5.85	122°	3.66
115°	−5.29	126°	4.43

dependence of *vicinal* ^{1}H–^{1}H couplings, the relationship between the conformation and *vicinal* couplings of, amongst others, ethane derivatives and peptide systems, has received much attention. The dihedral angle (φ) dependence of *vicinal* spin–spin couplings is similar to that given by the Karplus relation,

$$J^{vic} = A + B \cos \varphi + C \cos 2\varphi, \qquad (4.86)$$

where A, B and C are parameters which are determined empirically. *Vicinal* couplings have been a key factor in the conformational analysis of various systems and NMR spectroscopy has thus been used for determining conformational states, the observations being made in solutions appropriate for NMR studies.

The values of the parameters A, B and C for various couplings, obtained by the FPT-INDO procedure, are shown in Table 4.8.[14] The curves of $^3J(^1$H–^{1}H) against φ for the molecules considered resemble each other closely. The maxima occur at $\varphi = 180°$ and the minima are displaced somewhat from the 90° and 270° values of φ suggested by Karplus. Maciel *et al.*[42] have presented a general discussion of substituent effects on *vicinal* ^{1}H—^{1}H couplings in ethyl compounds. One tendency is that the substitution of an electronegative atom causes a decrease in magnitude in the average coupling.

Table 4.8 Values, in hertz, of the parameters A, B and C used in the Karplus relation together with FPT-INDO calculations

Molecules	A	B	C
Ethane			
H–C–C–H	8.436	−2.835	7.523
trans-N-Methylformamide			
H–N–C–H	4.899	−4.294	5.080
H–C–N–C	1.715	−1.546	2.336

(c) *Through-space Coupling*

The interaction between nuclei which are separated by several chemical bonds often arises from a through-space effect. A semi-empirical approach to the problem of through-space coupling is provided by equation (4.87):[45]

$$J(A-B) = \text{constant} \times S_A(0)^2 S_B(0)^2 S_{S_A S_B}^2, \qquad (4.87)$$

where $S_{S_A S_B}$ is the overlap integral between the valence s orbitals of atoms A and B, and the constant is determined experimentally.

Although equation (4.87) may be suitable for $^1H-^1H$ interactions, it does not appear to account satisfactorily for $^{19}F-^{19}F$ through-space contributions which depend upon the overlap of lone-pair p orbitals.[46] A large quantity of data for $^{19}F-^{19}F$ couplings are available, and a plot of $J(^{19}F-^{19}F)$ versus internuclear distance shows an excellent correlation with experiment for values of the internuclear separation in excess of 0.22 nm.

In the case of $^{19}F-^1H$ through-space couplings, both positive and negative contributions are found. These cannot be accounted for by equation (4.87), where the sign of the constant is determined by the product of the magnetogyric ratios of the interacting nuclei. The $^{19}F-^1H$ couplings in question may be rationalized by means of calculations of the contact term which shows the presence of several large contributions of either sign. Since the total contact interaction is obtained by summing these contributions its sign will depend on the occurrence of conformational averaging.

Rather good agreement with alkali metal halide spin–spin couplings is also obtained by using equation (4.87).

4.E MEDIUM EFFECTS ON SPIN–SPIN COUPLINGS

In the absence of specific effects the solvent dependence of coupling interactions is related to the polarity of the medium. This dependence has been studied theoretically by means of various approaches.[47-50] The details of these calculations are reviewed elsewhere.[44,51] Here the solvaton theory, which has been successfully used to interpret medium effects on nuclear shielding, it considered.

This theory was described in Section 3.D. It has been employed within the FPT formalism in an attempt to understand the effects of solvents on $^1J(^{13}C-^1H)$, $^3J(^1H-^1H)$ and $^3J(^{13}C-^1H)$. The approach adopted considers only the contact interaction. The spin-unrestricted SCF equations of the Fock matrices, $F_{\mu\nu}^\alpha$ and $F_{\mu\nu}^\beta$, for α and β electron spins become

$$F_{\mu\nu}^\alpha = \mathscr{H}_{\mu\nu}^{core} + \sum_\lambda \sum_\sigma [P_{\lambda\sigma}\langle\mu\nu|\lambda\sigma\rangle - P_{\lambda\sigma}^\alpha\langle\mu\sigma|\lambda\nu\rangle]$$
$$+ \frac{\varepsilon - 1}{\varepsilon} \int \chi_\mu^* \sum_s^n \frac{Q_s^\alpha}{r_{sl}} \chi_\nu \, dv + \frac{2\mu_0\mu_B\hbar}{3} \int \chi_\mu^* \delta(\mathbf{r}_N)\chi_\nu \, dv \quad (4.88)$$

and

$$F_{\mu\nu}^\beta = \mathscr{H}_{\mu\nu}^{core} + \sum_\lambda \sum_\sigma [P_{\lambda\sigma}\langle\mu\nu|\lambda\sigma\rangle - P_{\lambda\sigma}^\beta\langle\mu\sigma|\lambda\nu\rangle]$$
$$+ \frac{\varepsilon - 1}{\varepsilon} \int \chi_\mu^* \sum_s^n \frac{Q_s^\beta}{r_{sl}} \chi_\nu \, dv - \frac{2\mu_0\mu_B\hbar}{3} \int \chi_\mu^* \delta(\mathbf{r}_N)\chi_\nu \, dv, \quad (4.89)$$

where the fourth term in equations (4.88) and (4.89) represents the contact interaction and $\delta(\mathbf{r}_N)$ is the Dirac delta function which describes the contact interaction betweens electrons and the nucleus N; the other symbols have their usual meaning. The charges Q_s^α and Q_s^β for the α and β spins, respectively, on the solvaton are assumed to be equal in magnitude but opposite in sign to those on the atom to which it is considered to be attached. The spin–spin coupling is calculated by means of the FPT method incorporating the solute–solvent interaction into the Fock matrices given by equations (4.54) and (4.55).

The effects of solvents on the directly bonded ^{13}C–^1H spin–spin couplings of acrylonitrile, dichloromethane, chloroform, difluoromethane and trifluoromethane have been estimated by this procedure.[44,50] The calculations were performed within the INDO approximation for acrylonitrile, difluoromethane and chloroform and within the CNDO/2 approximation for dichloromethane and chloroform. The calculated values of $^1J(^{13}C$–$^1H)$ are presented in Table 4.9, as a function of ε, together with some experimental data. In all cases the calculated values are larger than the observed ones. However, the tendency for the calculated values to increase, with increasing ε, agrees qualitatively with the observed tendency.

The values of $^1J(^{13}C$–$^1H)$ produced by the solvaton model can be fitted approximately by

$$^1J(^{13}C\text{–}^1H) = a\left(\frac{\varepsilon - 1}{\varepsilon}\right) + b\left(\frac{\varepsilon - 1}{\varepsilon}\right)^2 + c. \quad (4.90)$$

The observed values of $^1J(^{13}C$–$^1H)$ fall approximately on a straight line when plotted against $(\varepsilon - 1)/\varepsilon$, consequently the second term in equation (4.90) can be neglected.

Table 4.9 ε dependences of the parameters A, B and C, in hertz, of the Karplus expression for ethane, $CHCl_2CHCl_2$ and trans-N-methylformamide, calculated by means of the FPT-solvaton method

Compound			Dielectric constant, ε						Coefficients in equation (4.91)	
			1	2	4	10	20	50	a'	b'
Ethane (H–C–C–H)										
	CNDO/2	A	6.835	6.985	6.986	6.997	6.997	6.997	0.18	6.835
		B	−1.869	−1.918	−1.919	−1.922	−1.922	−1.222	0.059	−1.869
		C	6.865	7.028	7.029	7.041	7.041	7.041	0.195	6.865
	INDO	A	8.346	8.445	8.493	8.516	8.516	8.516	0.189	8.346
		B	−2.835	−2.870	−2.889	−2.890	−2.291	−2.291	0.061	−2.835
		C	7.523	7.590	7.628	7.628	7.628	7.627	0.116	7.523
$CHCl_2CHCl_2$ (H–C–C–H)										
	CNDO/2	A	5.150	5.223	5.259	5.281	5.288	5.293	0.145	5.150
		B	−1.189	−1.223	−1.241	−1.253	−1.256	−1.258	0.071	−1.189
		C	5.324	5.413	5.456	5.483	5.493	5.500	0.079	5.324
trans-N-Methylformamide (H–N–C–H)										
	CNDO/2	A	4.315	4.333	4.343	4.349	4.351	4.352	0.038	4.315
		B	−2.722	−2.743	−2.745	−2.778	−2.781	−2.782	0.062	−2.722
		C	4.131	4.153	4.164	4.169	4.172	4.174	0.042	4.131
	INDO	A	4.899	4.988	4.963	4.974	4.978	4.981	0.083	4.899
		B	−4.294	−4.360	−4.393	−4.413	−4.420	−4.435	0.132	−4.294
		C	5.080	5.124	5.149	5.164	5.169	5.172	0.093	5.080
trans-N-Methylformamide (H–C–N–C)										
	CNDO/2	A	1.650	1.668	1.677	1.683	1.685	1.686	0.036	1.650
		B	−0.998	−1.016	−1.026	−1.032	−1.035	−1.036	0.038	−0.998
		C	1.597	1.598	1.599	1.604	1.605	1.607	0.009	1.597
	INDO	A	1.715	1.765	1.793	1.814	1.822	1.827	0.110	1.715
		B	−1.546	−1.588	−1.611	−1.626	−1.630	−1.634	0.089	−1.546
		C	2.336	2.335	2.335	2.333	2.331	2.330	−0.004	2.336

The SOS-INDO method, incorporating the solvaton model, has been employed to obtain the variation of $^1J(^{15}N-^{13}C)$ as a function of ε for some cyanides and isocyanides.[52] In all cases considered $^1J(^{15}N-^{13}C)$ is predicted to become increasingly negative as ε increases. Changes of up to 2 Hz in $^1J(^{15}N-^{13}C)$ are expected, with the isocyanides being more sensitive than the cyanides to a change in ε. This is consistent with the more polar structure usually considered for isocyanides.

In molecules which can adopt two or more conformations, two effects on the couplings must be considered, one due to solvent-induced electronic changes in the solute molecules and the other arising from a change in the relative populations between the conformers on account of intermolecular solute–solvent interactions. The former effect in particular must be considered here in order to clarify the above problem. As described in Section 4.D.6, the dihedral angle, φ, dependence of *vicinal* spin–spin couplings in an isolated molecule is given by the Karplus equation (4.86). In order to provide a more exact insight into the relationship between the magnitude of the *vicinal* coupling and conformation, solvent effects have to be taken into account. A theoretical prediction of solvent effects on *vicinal* proton–proton and proton–carbon couplings has been provided.

The values of the coefficients A, B and C for ethane, $CHCl_2CHCl_2$ and *trans-N*-methylformamide, have been calculated as a function of ε by means of the FPT-INDO and FPT-CNDO/2 approximations within the solvaton model. The absolute values of the coefficients A, B and C in the CNDO/2 calculations are larger than those found from INDO calculations for a range of dielectric constants. The values of the coefficients obtained for the H–C–C–H couplings in ethanes decrease upon substitution of chlorine for hydrogen, in the CNDO calculations. For the H–C–N–H couplings the values are smaller than those appropriate to the H–C–C–H and H–C–N–H couplings. The value of B in the former is somewhat larger than in the latter case. The calculated absolute values of the coefficients A, B and C show an increase with increasing ε except in the case of the coefficient C for $^3J(^{13}C-^1H)$ in *trans-N*-methylformamide obtained from an INDO calculation. The values of A, B and C show the following dependence on ε:

$$A, B \text{ or } C = a'\left(\frac{\varepsilon - 1}{\varepsilon}\right) + b', \tag{4.91}$$

where a' is a measure of the strength of the dielectric solvent effect on the coefficients A, B and C and b' is the value of A, B or C at $\varepsilon = 1$, the appropriate values of a' and b' are indicated in the last two columns of Table 4.9.

In conclusion, we find that approximate theoretical models of medium effects on spin–spin couplings are reasonable as a means of describing nonbonded solute–solvent interactions. Obviously, there is much room for improvement. The semi-empirical MO descriptions of spin–spin coupling for isolated molecules require improvement in the details of the approximations involved and in the parameter sets in current use. Bearing these imperfections in mind it is hardly surprising that, with the additional approximations in the solute–solvent interaction model, the overall agreement between observed and predicted solvation effects on NMR parameters is less than quantitatively satisfactory in many cases at the present time.

REFERENCES

1. M. Barfield and D. M. Grant, in "Advances in Magnetic Resonance", Vol. 1 (J. S. Waugh, ed.), Academic Press, New York (1965), p. 149.
2. J. N. Murrell, in "Progress in NMR Spectroscopy", Vol. 6 (J. W. Emsley, J. Feeney and L. H. Sutcliffe, eds), Pergamon Press, Oxford (1970), p. 1.
3. P. D. Ellis and R. Ditchfield, in "Topics in ^{13}C NMR Spectroscopy", Vol. 2 (G. C. Levy, ed.), John Wiley & Sons, New York (1976), p. 433.
4. J. Kowalewski, in "Progress in NMR Spectroscopy", Vol. 11 (J. W. Emsley, J. Feeney and L. H. Sutcliffe, eds), Pergamon Press, Oxford (1977), p. 1.
5. J. Kowalewski, in "Annual Reports on NMR Spectroscopy", Vol. 12 (G. A. Webb, ed.), Academic Press, London (1982), p. 81.
6. H. S. Gutowsky and D. W. McCall, *Phys. Rev.*, **82**, 748 (1951).
7. H. S. Gutowsky, D. W. McCall and C. P. Slichter, *Phys. Rev.*, **84**, 589 (1951).
8. E. L. Hahn and D. E. Maxwell, *Phys. Rev.*, **84**, 1246 (1951).
9. N. F. Ramsey and E. M. Purcell, *Phys. Rev.*, **85**, 143 (1952).
10. N. F. Ramsey, *Phys. Rev.*, **91**, 303 (1953).
11. H. Y. Carr and E. M. Purcell, *Phys. Rev.*, **88**, 415 (1952).
12. J. A. Pople and D. P. Santry, *Molec. Phys.*, **8**, 1 (1964).
13. J. A. Pople, J. W. McIver and N. S. Ostlund, *J. Chem. Phys.*, **49**, 2960 (1968).
14. A. C. Blizzard and D. P. Santry, *J. Chem. Phys.*, **55**, 950 (1971).
15. M. Karplus, D. H. Anderson, T. C. Farrar and H. S. Gutowsky, *J. Chem. Phys.*, **27**, 597 (1957).
16. M. Karplus and D. H. Anderson, *J. Chem. Phys.*, **30**, 6 (1959).
17. G. Rumer, *Nachr. ges. Wiss. Göttingen*, **1932**, 337 (1932).
18. L. Pauling, *J. Chem. Phys.*, **1**, 280 (1933).
19. M. Barfield, *J. Chem. Phys.*, **46**, 811 (1967).
20. M. Barfield, *J. Chem. Phys.*, **48**, 4458 (1968).
21. R. Ditchfield, N. S. Ostlund, J. N. Murrell and M. A. Turpin, *Molec. Phys.*, **18**, 433 (1970).
22. J. Kowalewski, R. Vestin and B. Roos, *Chem. Phys. Letters*, **12**, 25 (1971).
23. H. Benoit and P. Piejus, *C. r. hebd. Séanc. Acad. Sci., Paris, B*, **265**, 101 (1967).
24. Y. Kato and A. Saika, *J. Chem. Phys.*, **46**, 1975 (1967).
25. J. S. Muenter and W. Klemperer, *J. Chem. Phys.*, **52**, 6033 (1970).
26. A. D. C. Towl and K. Schaumburg, *Molec. Phys.*, **22**, 49 (1971).
27. H. Nakatuji, K. Hirao, H. Kato and T. Yonezawa, *Chem. Phys. Letters*, **6**, 541 (1970).

28. T. Khin and G. A. Webb, *Org. magn. Reson.*, **10**, 175 (1977).
29. T. Khin and G. A. Webb, *Org. magn. Reson.*, **11**, 487 (1978).
30. A. Laaksonen and J. Kowalewski, *J. Am. chem. Soc.*, **103**, 5277 (1981).
31. J. A. Pople, J. W. McIver and N. S. Ostlund, *Chem. Phys. Letters*, **1**, 465 (1967).
32. G. E. Maciel, J. W. McIver, N. S. Ostlund and J. A. Pople, *J. Am. chem. Soc.*, **92**, 1 (1970).
33. M. J. S. Dewar, D. Landman, Sung Ho Suck and P. K. Weiner, *J. Am. chem. Soc.*, **99**, 3951 (1977).
34. W. S. Lee and J. M. Schulman, *J. Am. chem. Soc.*, **101**, 3182 (1979).
35. A. A. Cheremisin and P. V. Schastnev, *Zh. strukt. Khim.*, **21**, 177 (1980).
36. H. Günther, H. Seel and M.-E. Günther, *Org. magn. Reson.*, **11**, 97 (1978).
37. J. M. Schulman and T. Venanzi, *J. Am. chem. Soc.*, **98**, 6739 (1976).
38. M. D. Newton and J. N. Schulman, *J. Am. chem. Soc.*, **94**, 767, 773 (1972).
39. M. D. Newton, J. M. Schulman and M. M. Manus, *J. Am. chem. Soc.*, **96**, 17 (1974).
40. J. M. Schulman and M. D. Newton, *J. Am. chem. Soc.*, **96**, 6295 (1974).
41. N. Müller and D. E. Pritchard, *J. Chem. Phys.*, **31**, 768 (1959).
42. G. E. Maciel, J. W. McIver, N. S. Ostlund and J. A. Pople, *J. Am. chem. Soc.*, **92**, 4151 (1970).
43. M. Karplus, *J. Chem. Phys.*, **30**, 11 (1959).
44. I. Ando and G. A. Webb, *Org. magn. Reson.*, **15**, 111 (1981).
45. A. D. Buckingham and J. E. Cordle, *J. chem. Soc. Faraday II*, **70**, 994 (1974).
46. J. Hilton and L. H. Sutcliffe, *J. chem. Soc. Faraday II*, **71**, 1395 (1975).
47. M. D. Johnston and M. Barfield, *J. Chem. Phys.*, **54**, 3083 (1971).
48. M. D. Johnston and M. Barfield, *J. Chem. Phys.*, **55**, 3483 (1971).
49. M. D. Johnston and M. Barfield, *Molec. Phys.*, **22**, 831 (1971).
50. M. Kondo, S. Watanabe and I. Ando, *Molec. Phys.*, **37**, 1521 (1979).
51. M. Barfield and M. D. Johnston, *Chem. Rev.*, **73**, 53 (1973).
52. S. N. Shargi and G. A. Webb, *Org. magn. Reson.*, **19**, 126 (1982).

5
Program Listings

5.A A COMPUTER PROGRAM FOR CALCULATING NUCLEAR SHIELDING BY MEANS OF FPT-INDO AND CNDO/2 METHODS

A program written for the HITAC M200 computer is presented below for calculating nuclear shielding by means of the FPT-INDO and CNDO/2 methods. The program is suitable for calculating nuclear shielding for some first row elements, B, C, N, O and F. It is based on the program for CNDO and INDO calculations available from the Quantum Chemistry Program Exchange, Department of Chemistry, Indiana University, Bloomington, Indiana, USA, which was modified to incorporate nuclear shielding.

A paper by Kondo et al.[1] should be consulted for the details of the FPT calculation on nuclear shielding. In this program the bonding parameters for carbon and hydrogen atoms, β_C and β_H, are revised from those given by Pople[2] from $\beta_C = -21$ eV and $\beta_H = -9$ eV to $\beta_C = -15$ eV and $\beta_H = -13$ eV. The latter parameters provide good agreement between the observed and calculated values for the shielding of carbon atoms in hydrocarbon compounds. The bonding parameters used for the other atoms are those presented by Pople.

5.A.1 Description of the Program

In the main program, input data for a calculation is read. After reading the input data, the main program calls the subroutines COEFFT and INTEGRAL in which the integrals for the MO calculation are computed. The computation is carried out for the molecular electronic energy which is subjected to perturbations due to interaction between the electrons and the applied magnetic field (HUCKCL, SCFCLO). The descriptions of the subroutines COEFFT, INTEGL and SCFLO are substantially the same as those given in Pople's program for CNDO and INDO calculations.[3] The subroutine CHSFT calculates the nuclear shielding as a sum of paramagnetic and diamagnetic contributions. The paramagnetic shielding is estimated as either a monoatomic contribution or a sum of monoatomic and neighbouring atom contributions.

In the program, the maximum number of atoms allowed in the molecules concerned is 35 and the maximum number of orbitals is 80.

5.A.2 Description of the Input Data

First card   NAME   FORMAT (20A4)
Alphanumeric information to be printed on top of output sheet.

Second card   OPTION, OPNCLO, CONTR
    FORMAT (A4, 1X, A4, 1X, A4)
OPTION: Either "CNDO" or, "INDO" should be chosen.
OPNCLO: Either "OPEN" or "CLSD" should be chosen.
CONTR: Either "MONO" or "NEIG" should be chosen. "MONO" means that the paramagnetic shielding is estimated as a monoatomic contribution and "NEIG" that the paramagnetic shielding is estimated as a sum of monoatomic and neighbouring atom contributions.

Third card   NATOMS, CHARGE, MULTIP   FORMAT(3I4)
NATOMS: Number of atoms in the molecule.
CHARGE: Charge of the molecule.
MULTIP: Multiplicity.

Next cards   AN, C   FORMAT(I4, 6X, 3F10.5)
AN: Atomic number.
C: Cartesian coordinates of atoms in the molecule.

5.A.3 The Program

```
C
C       A PROGRAM FOR CALCULATION OF SOME FIRST ROW NUCLEI ( BORON,
C       CARBON,NITROGEN,OXYGEN AND FLUORINE ) BY MEANS OF FINITE
C       PERTURBATION METHOD
C
      IMPLICIT REAL*8 (A-H,O-Z)
      COMMON/ARRAYS/ABC(19200)
      COMMON/INFO/NATOMS,CHARGE,MULTIP,AN(35),C(35,3),N
      COMMON/PERTBL/EL(18),NAME(20)
      COMMON/ORB/ORB(9)
      COMMON/GAB/XYZ(2000)
      COMMON/INFO1/CZ(35),U(80),ULIM(35),LLIM(35),NELECS,OCCA,OCCB
```

5 PROGRAM LISTINGS

```
      COMMON/OPTION/OPTION,OPNCLO,HUCKEL,CNDO,INDO,CLOSED,OPEN,CONTR,MON
      COMMON/AUXINT/A(17),B(17)
      COMMON/IMAG/AIMAGE(12880)
      COMMON/COMP/COMP(6400),BGAUSS,IER
      COMMON/EORK/WORK1(80),WORK2(12800),LAMBDA(80),IWORK(160)
      COMMON/GAUSTR/V1S2P(100),V2S2P(100),V2P2S(100),V2P2P1(100),V2P2P2(
     1 100),T1S2P(100),T2S2P(100),T2P2P1(100),T2P2P2(100),U2S2P(100),U2P
     2 2P1(100),U2P2P2(100),INDX1(35,35),INDX2(35,35),INVJ1I(100),INVJ1J
     3 (100),INVJ2I(100),INVJ2J(100),INVT1I(100),INVT1J(100),INVT2I(100)
     4 ,INVT2J(100),DXX(35,35),Y1(80,3,3),Y2(80,3,3)
      COMPLEX*16 COMP,LAMBDA,WORK1,WORK2
      INTEGER OPTION,OPNCLO,HUCKEL,CNDO,INDO,CLOSED,OPEN,CONTR,MONO,NEIG
      INTEGER ORB,EL,AN,CHARGE,CZ,U,ULIM,OCCA,OCCB
  100 CONTINUE
      READ(5,20,END=99)(NAME(I),I=1,20)
   20 FORMAT(20A4)
      WRITE(6,30) (NAME(I),I=1,20)
   30 FORMAT(1H1,20A4)
      READ(5,40)OPTION,OPNCLO,CONTR
   40 FORMAT(A4,1X,A4,1X,A4)
      WRITE(6,45)OPTION,OPNCLO,CONTR
   45 FORMAT(1H0,10X,2A4,'METHOD',1X,A4,'CONTRIBUTION')
      READ(5,50) NATOMS,CHARGE,MULTIP
   50 FORMAT(3I4)
      WRITE(6,55) NATOMS,CHARGE,MULTIP
   55 FORMAT(1H0,5X,I4,' ATOMS    CHARGE =',I4,'    MULTIPLICITY =',I4/)
      DO 10 I=1,NATOMS
      READ(5,70)AN(I),C(I,1),C(I,2),C(I,3)
   70 FORMAT(I4,6X,3F10.5)
      WRITE(6,75) AN(I),C(I,1),C(I,2),C(I,3)
   75 FORMAT(1H ,10X,I4,6X,3F10.5)
      DO 9 J=1,3
    9 C(I,J)=C(I,J)/0.529167D0
   10 CONTINUE
      BGAUSS=500.
      WRITE(6,65) BGAUSS
   65 FORMAT(1H0,5X,'EXTERNAL MAGNETIC FIELD = ',F10.4,' X 10**4 GAUSS')
      IF (OPTION.EQ.CNDO) GO TO 6
    1 DO 5 I=1,NATOMS
      IF(AN(I).LE.9) GO TO 4
    2 WRITE(6,3)
    3 FORMAT(1H0,5X,'THIS PROGRAM DOES NOT DO INDO CALCULATIONS FOR MOLE
     1CULES CONTAINED ELEMENTS HIGHER THAN FLUORINE')
      STOP
    4 CONTINUE
    5 CONTINUE
    6 CONTINUE
      CALL COEFFT
      CALL INTGRL
      LLL=0
  200 CONTINUE
      IF(LLL.GE.3) GO TO 300
      LLL=LLL+1
      CALL HUCKCL(LLL)
      CALL SCFCLO
      CALL CHSFT(LLL)
      GO TO 200
  300 CONTINUE
      GO TO 100
```

```
   99 CONTINUE
      CALL EXIT
      STOP
      END



      BLOCK DATA
      COMMON/ORB/ORB(9)
      COMMON/PERTBL/EL(18),NAME(20)
      COMMON/OPTION/OPTION,OPNCLO,HUCKEL,CNDO,INDO,CLOSED,OPEN,CONTR,MON
     10,NEIG
      INTEGER OPTION,OPNCLO,HUCKEL,CNDO,INDO,CLOSED,OPEN,CONTR,MONO,NEIG
      INTEGER ORB,EL
      DATA CNDO/'CNDO'/
      DATA INDO/'INDO'/
      DATA OPEN/'OPEN'/
      DATA CLOSED/'CLSD'/
      DATA MONO/'MONO'/
      DATA NEIG/'NEIG'/
      DATA ORB/'    S',' PX',' PY',' PZ',' DZ2',' DXZ',' DYZ','DX-Y',
     1 ' DXY'/
      DATA EL/'   H','  HE','  LI','  BE','   B','   C','   N','   O',
     1 '   F','  NE','  NA','  MG','  AL',' SI','   P','   S','  CL',
     2 '  AR'/
      END


      SUBROUTINE INTORL
      IMPLICIT REAL*8(A-H,O-Z)
C     ATOMIC INTEGRALS FOR CNDO CALCULATIONS
      COMMON/ARRAYS/S(80,80),Y(9,5,203),Z(17,45),XX(2900)
      COMMON/INFO/NATOMS,CHARGE,MULTIP,AN(35),C(35,3),N
      COMMON/INFO1/CZ(35),U(80),ULIM(35),LLIM(35),NELECS,OCCA,OCCB
      COMMON/GAB/XXX(400),GAMMA(35,35),T(9,9),PAIRS(9,9),TEMP(9,9),C1(3)
     1,C2(3),YYY(126)
      COMMON/AUXINT/A(17),B(17)
      COMMON/OPTION/OPTION,OPNCLO,HUCKEL,CNDO,INDO,CLOSED,OPEN,CONTR,MON
     10,NEIG
      COMMON/COMP/BCOMPL(80,80),BGAUSS,IER
      COMMON/GAUSTR/V1S2P(100),V2S2P(100),V2P2S(100),V2P2P1(100),V2P2P2(
     1 100),T1S2P(100),T2S2P(100),T2P2P1(100),T2P2P2(100),U2S2P(100),U2P
     2 2P1(100),U2P2P2(100),INDX1(35,35),INDX2(35,35),INVJ1I(100),INVJ1.
     3 (100),INVJ2I(100),INVJ2J(100),INVT1I(100),INVT1J(100),INVT2I(100
     4 ,INVT2J(100),D(35,35),Y1(80,3,3),Y2(80,3,3)
      COMMON/PERTBL/EL(18),NAME(20)
      DIMENSION APAIRS(9,9),ATEMP(9,9),AS(80,80)
      DIMENSION MU(18),NC(18),LC(9),MC(9),E(3)
      DIMENSION P(80,80)
      EQUIVALENCE (P(1),Y(1))
      EQUIVALENCE (APAIRS(1),XX(1)),(ATEMP(1),XX(101)),(AS(1),BCOMPL(1)
      EXTERNAL FUNC
      COMPLEX*16 BCOMPL
      REAL*8 MU,NUM,K1,K2
      INTEGER AN,ULIM,ULK,ULL,CZ,U,CHARGE,ANL,ANK,OCCA,OCCB,EL
      INTEGER OPTION,OPNCLO,HUCKEL,CNDO,INDO,CLOSED,OPEN,CONTR,MONO,NEI
      N=0
      DO 60 I=1,NATOMS
```

```
      LLIM(I)=N+1
      K=1
      IF(AN(I).LT.11) GO TO 20
   10 N=N+9
      CZ(I)=AN(I)-10
      GO TO 50
   20 IF(AN(I).LT.3) GO TO 40
   30 N=N+4
      CZ(I)=AN(I)-2
      GO TO 50
   40 N=N+1
      CZ(I)=AN(I)
   50 CONTINUE
      ULIM(I)=N
   60 CONTINUE
C     FILL U ARRAY----U(J) IDENTIFIES THE ATOM TO WHICH ORBITAL J IS
C     ATTACHED E.G. ORBITAL32 ATTACHED TO ATOM 7, ETC.
      DO 70 K=1,NATOMS
      LLK=LLIM(K)
      ULK=ULIM(K)
      LIM=ULK+1-LLK
      DO 70 I=1,LIM
      J=LLK+I-1
   70 U(J)=K
C     ASSIGNMENT OF ORBITAL EXPONENTS TO ATOMS BY SLATERS RULES
      MU(2)=1.7D0
      MU(1)=1.2D0
      NC(1)=1
      NC(2)=1
      DO 80 I=3,10
      NC(I)=2
   80 MU(I)=0.325D0*DFLOAT(I-1)
      DO 90 I=11,18
      NC(I)=3
   90 MU(I)=(0.65D0*DFLOAT(I)-4.95D0)/3.D0
      LC(1)=0
      LC(2)=1
      LC(3)=1
      LC(4)=1
      LC(5)=2
      LC(6)=2
      LC(7)=2
      LC(8)=2
      LC(9)=2
      MC(1)=0
      MC(2)=1
      MC(3)=-1
      MC(4)=0
      MC(5)=0
      MC(6)=1
      MC(7)=-1
      MC(8)=2
      MC(9)=-2
      BSIGMA=1.D0
      BPI=0.85D0
      DO 320 K=1,NATOMS
      DO 320 L=K,NATOMS
      DO 100 I=1,3
      C1(I)=C(K,I)
```

```
100 C2(I)=C(L,I)
    CALL RELVEC(R,E,C1,C2)
    LLK=LLIM(K)
    LLL=LLIM(L)
    ULK=ULIM(K)
    ULL=ULIM(L)
    NORBK=ULK-LLK+1
    NORBL=ULL-LLL+1
    ANK=AN(K)
    ANL=AN(L)
    DO 200 I=1,NORBK
    DO 200 J=1,NORBL
    IF(K.EQ.L) GO TO 160
    BCONST=BSIGMA
    IF((MC(I).EQ.O).OR.(MC(J).EQ.O)) GO TO 105
    IF(MC(I).EQ.MC(J)) BCONST=BPI
105 CONTINUE
110 IF(MC(I).NE.MC(J)) GO TO 150
120 IF (MC(I).LT.0) GO TO 140
130 PAIRS(I,J)=DSQRT((MU(ANK)*R)**(2*NC(ANK)+1)*(MU(ANL)*R)**(2*NC(ANL
   1)+1)/(FACT(2*NC(ANK))*FACT(2*NC(ANL))))*(-1.D0)**(LC(J)+MC(J))
   2*SS(NC(ANK),LC(I),MC(I),NC(ANL),LC(J),MU(ANK)*R,MU(ANL)*R)
    APAIRS(I,J)=PAIRS(I,J)*BCONST
    GO TO 190
140 PAIRS(I,J)=PAIRS(I-1,J-1)
    APAIRS(I,J)=APAIRS(I-1,J-1)
    GO TO 190
150 PAIRS(I,J)=0.0D0
    APAIRS(I,J)=0.0D0
    GO TO 190
160 IF(I.EQ.J) GO TO 170
180 PAIRS(I,J)=0.0D0
    APAIRS(I,J)=0.0D0
    GO TO 190
170 PAIRS(I,J)=1.0D0
    APAIRS(I,J)=1.0D0
190 CONTINUE
200 CONTINUE
    LCULK=LC(NORBK)
    LCULL=LC(NORBL)
    MAXL=MAX0(LCULK,LCULL)
    IF(R.GT.0.000001D0) GO TO 220
210 GO TO 250
220 CALL HARMTR(T,MAXL,E)
    DO 230 I=1,NORBK
    DO 230 J=1,NORBL
    TEMP(I,J)= 0.D0
    ATEMP(I,J)=0.0D0
    DO 230 KK=1,NORBL
    TEMP(I,J)=TEMP(I,J)+T(J,KK)*PAIRS(I,KK)
    ATEMP(I,J)=ATEMP(I,J)+T(J,KK)*APAIRS(I,KK)
230 CONTINUE
    DO 240 I=1,NORBK
    DO 240 J=1,NORBL
    PAIRS(I,J)=0.0D0
    APAIRS(I,J)=0.0D0
    DO 240 KK=1,NORBK
    PAIRS(I,J)=PAIRS(I,J)+T(I,KK)*TEMP(KK,J)
    APAIRS(I,J)=APAIRS(I,J)+T(I,KK)*ATEMP(KK,J)
```

```
  240 CONTINUE
C     FILL S MATRIX
  250 CONTINUE
      DO 260 I=1,NORBK
      LLKP=LLK+I-1
      DO 260 J=1,NORBL
      LLLP=LLL+J-1
      S(LLKP,LLLP)=PAIRS(I,J)
      AS(LLKP,LLLP)=APAIRS(I,J)
  260 CONTINUE
      N1=NC(ANK)
      N2=NC(ANL)
      K1=MU(ANK)
      K2=MU(ANL)
      IF(K.NE.L) GO TO 290
  270 TERM1=FACT(2*N1-1)/((2.D0*K2)**(2*N1))
      TERM2=0.D0
      LIM=2*N1
      DO 280 J=1,LIM
      NN1=4*N1-J
      NUM=DFLOAT(J)*(2.D0*K1)**(2*N1-J)*FACT(NN1-1)
      NN2=2*N1-J
      DEN=FACT(NN2)*2.D0*DFLOAT(N1)*(2.D0*(K1+K2))**(4*N1-J)
      TERM2=TERM2+NUM/DEN
  280 CONTINUE
      GO TO 310
  290 TERM1=(R/2.D0)**(2*N2)*SS(0,0,0,2*N2-1,0,0.D0,2.D0*K2*R)
      TERM2=0.D0
      LIM=2*N1
      DO 300 J=1,LIM
  300 TERM2=TERM2+(DFLOAT(J)*(2.D0*K1)**(2*N1-J)*(R/2.D0)**(2*
     1N1-J+2*N2))/(FACT(2*N1-J)*2.D0*DFLOAT(N1))*SS(2*N1-J,0,0,2*N2-1,0
     2,2.D0*K1*R,2.D0*K2*R)
  310 GAMMA(K,L)=((2.D0*K2)**(2*N2+1)/FACT(2*N2))*(TERM1-TERM2)
  320 CONTINUE
      DO 330 I=1,N
      DO 330 J=I,N
      AS(J,I)=AS(I,J)
  330 S(J,I)=S(I,J)
      DO 340 I=1,NATOMS
      DO 340 J=I,NATOMS
  340 GAMMA(J,I)=GAMMA(I,J)
      WRITE(6,350)
  350 FORMAT(1H1,1X,23HOVERLAP INTEGRAL MATRIX)
      CALL MATOUT(N,1)
      DO 360 I=1,NATOMS
      DO 360 J=1,NATOMS
  360 P(I,J)=GAMMA(I,J)
      WRITE(6,370)
  370 FORMAT(1X,23HCOULOMB INTEGRAL MATRIX)
      CALL MATOUT(NATOMS,2)
      DO 380 I=1,N
      DO 380 J=I,N
      S(I,J)=AS(I,J)
      S(J,I)=S(I,J)
  380 CONTINUE
      IF(CONTR.EQ.MONO)RETURN
      DO 32 I=1,NATOMS
      D(I,I)=0.D0
```

```
      DO 32 J=1,NATOMS
      INDX1(I,J)=0
      INDX2(I,J)=0
   32 CONTINUE
      DO 33 I=2,NATOMS
      DO 33 J=1,I-1
      D(I,J)=(C(I,1)-C(J,1))**2+(C(I,2)-C(J,2))**2+(C(I,3)-C(J,3))**2
      D(J,I)=D(I,J)
   33 CONTINUE
      C0=0.D0
      D0=1.D0
      E0=0.D0
      F0=1.D0
      INTJ1=0
      INTJ2=0
      DO 18 I=2,NATOMS
      IF(AN(I).LT.3) GO TO 19
      DO 21 J=1,I-1
      IF(AN(J).LT.3) GO TO 22
      DO 23 K=2,I
      IF(AN(K).LT.3) GO TO 23
      DO 7223 L=1,K-1
      IF((AN(I).NE.AN(K)).AND.(AN(I).NE.AN(L))) GO TO 7223
      IANDJ=AN(I)+AN(J)
      KANDL=AN(K)+AN(L)
      IF(IANDJ.NE.KANDL) GO TO 7223
      DIF=DABS(D(I,J)-D(K,L))
      IF(DIF.GT.0.000001D0) GO TO 7223
      INDX1(I,J)=IND(1(K,L)
      IF((I.EQ.K).AND.(J.EQ.L)) GO TO 24
      GO TO 21
 7223 CONTINUE
   23 CONTINUE
   24 INTJ1=INTJ1+1
      INDX1(I,J)=INTJ1
      INVJ1I(INTJ1)=I
      INVJ1J(INTJ1)=J
      GO TO 21
   22 CONTINUE
      DO 25 K=2,I
      DO 25 L=1,K-1
      IF((AN(I).NE.AN(K)).AND.(AN(I).NE.AN(L))) GO TO 25
      IANDJ=AN(I)+AN(J)
      KANDL=AN(K)+AN(L)
      IF(IANDJ.NE.KANDL) GO TO 25
      DIF=DABS(D(I,J)-D(K,L))
      IF(DIF.GT.0.000001D0) GO TO 25
      INDX1(I,J)=INDX1(K,L)
      IF((I.EQ.K).AND.(J.EQ.L)) GO TO 26
      GO TO 21
   25 CONTINUE
   26 CONTINUE
      INTJ2=INTJ2+1
      INDX1(I,J)=INTJ2
      INVJ2I(INTJ2)=I
      INVJ2J(INTJ2)=J
   21 CONTINUE
      GO TO 18
   19 CONTINUE
```

5 PROGRAM LISTINGS

```
      DO 27 J=1,I-1
      IF(AN(J).LT.3) GO TO 27
      DO 28 K=2,I
      DO 28 L=1,K-1
      IF((AN(I).NE.AN(K)).AND.(AN(I).NE.AN(L))) GO TO 28
      IANDJ=AN(I)+AN(J)
      KANDL=AN(K)+AN(L)
      IF(IANDJ.NE.KANDL) GO TO 28
      DIF=DABS(D(I,J)-D(K,L))
      IF(DIF.GT.0.000001D0) GO TO 28
      INDX1(I,J)=INDX1(K,L)
      IF((I.EQ.K).AND.(J.EQ.L)) GO TO 29
   28 CONTINUE
   29 CONTINUE
      INTJ2=INTJ2+1
      INDX1(I,J)=INTJ2
      INVJ2I(INTJ2)=I
      INVJ2J(INTJ2)=J
   27 CONTINUE
   18 CONTINUE
      DO 520 I=1,INTJ2
      ASQ=D(INVJ2I(I),INVJ2J(I))
      ALPHA1=MU(AN(INVJ2I(I)))
      ALPHA2=MU(AN(INVJ2J(I)))
      ALPMUL=ALPHA1*ALPHA2
      YY0=DGASPD(FUNC,6,6,C0,D0,E0,F0,1,3,0,1,ASQ,ALPHA1,ALPHA2)
      VIS2P(I)=-0.125D0*ALPMUL**2.5*ALPHA2*YY0
  520 CONTINUE
      DO 530 I=1,INTJ1
      ASQ=D(INVJ1I(I),INVJ1J(I))
      ALPHA1=MU(AN(INVJ1I(I)))
      ALPHA2=MU(AN(INVJ1J(I)))
      ALPMUL=ALPHA1*ALPHA2
      YY0=DGASPD(FUNC,6,6,C0,D0,E0,F0,1,3,0,0,ASQ,ALPHA1,ALPHA2)
      YY1=DGASPD(FUNC,6,6,C0,D0,E0,F0,1,3,1,0,ASQ,ALPHA1,ALPHA2)
      YY2=DGASPD(FUNC,6,6,C0,D0,E0,F0,1,3,0,1,ASQ,ALPHA1,ALPHA2)
      YY3=DGASPD(FUNC,6,6,C0,D0,E0,F0,1,4,1,1,ASQ,ALPHA1,ALPHA2)
      V2S2P(I)=0.0721687837D0*ALPMUL**2.5*ALPHA2*
     1 (YY2-0.5D0*ALPHA1**2*YY3)
      V2P2S(I)=0.0721687837D0*ALPMUL**2.5*ALPHA1*
     1 (YY1-0.5D0*ALPHA2**2*YY3)
      V2P2P1(I)=0.125D0*ALPMUL**3.5*YY3
      V2P2P2(I)=-0.125D0*ALPMUL**3.5*YY0
  530 CONTINUE
      INTT1=0
      INTT2=0
      DO 11 I=1,NATOMS
      IF(AN(I).NE.6) GO TO 11
      DO 12 J=1,NATOMS
      IF(I.EQ.J) GO TO 12
      IF(AN(J).LT.3) GO TO 13
      DO 7214 K=1,I
      IF(AN(K).NE.6) GO TO 7214
      DO 14 L=1,NATOMS
      IF(AN(J).NE.AN(L)) GO TO 14
      DIF=DABS(D(I,J)-D(K,L))
      IF(DIF.GT.0.000001D0) GO TO 14
      INDX2(I,J)=INDX2(K,L)
      IF((I.EQ.K).AND.(J.EQ.L)) GO TO 15
```

```
      GO TO 12
   14 CONTINUE
 7214 CONTINUE
   15 CONTINUE
      INTT1=INTT1+1
      INDX2(I,J)=INTT1
      INVT1I(INTT1)=I
      INVT1J(INTT1)=J
      GO TO 12
   13 CONTINUE
      DO 16 K=1,I
      IF(AN(K).NE.6) GO TO 7216
      DO 16 L=1,NATOMS
      IF(AN(L).GT.2) GO TO 16
      DIF=DABS(D(I,J)-D(K,L))
      IF(DIF.GT.0.000001D0) GO TO 16
      INDX2(I,J)=INDX2(K,L)
      IF((I.EQ.K).AND.(J.EQ.L)) GO TO 17
      GO TO 12
   16 CONTINUE
 7216 CONTINUE
   17 CONTINUE
      INTT2=INTT2+1
      INDX2(I,J)=INTT2
      INVT2I(INTT2)=I
      INVT2J(INTT2)=J
   12 CONTINUE
   11 CONTINUE
      IND=2
      M1=4
      M2=4
      IF(INTT2.EQ.0) GO TO 498
      DO 500 I=1,INTT2
      ASQ=D(INVT2I(I),INVT2J(I))
      ALPHA1=MU(AN(INVT2J(I)))
      ALPHA2=MU(6)
      ALPMUL=ALPHA1*ALPHA2
      YY0=DGASPD(FUNC,M1,M2,C0,D0,E0,F0,IND,1,-2,1,ASQ,ALPHA1,ALPHA2)
      T1S2P(I)=0.1591549431D0*ALPMUL**2.5*ALPHA2*ASQ*YY0
  500 CONTINUE
  498 CONTINUE
      IF(INTT1.EQ.0) GO TO 499
      DO 510 I=1,INTT1
      ASQ=D(INVT1I(I),INVT1J(I))
      ALPHA1=MU(AN(INVT1J(I)))
      ALPHA2=MU(6)
      ALPMUL=ALPHA1*ALPHA2
      YY0=DGASPD(FUNC,M1,M2,C0,D0,E0,F0,IND,1,-2,1,ASQ,ALPHA1,ALPHA2)
      YY1=DGASPD(FUNC,M1,M2,C0,D0,E0,F0,IND,2,-3,1,ASQ,ALPHA1,ALPHA2)
      YY2=DGASPD(FUNC,M1,M2,C0,D0,E0,F0,IND,3,-4,1,ASQ,ALPHA1,ALPHA2)
      YY3=DGASPD(FUNC,M1,M2,C0,D0,E0,F0,IND,3,-5,2,ASQ,ALPHA1,ALPHA2)
      T2S2P(I)=-0.09188814924D0*ALPMUL**2.5*ALPHA2*ASQ*(YY0-0.5D0*
     1 ALPHA1**2*ASQ*YY1)
      T2P2P1(I)=0.3183098862D0*ALPMUL**3.5*ASQ*(YY1-0.125D0*ASQ*
     1 (ALPHA1**2*YY2+ALPHA2**2*YY3))
      T2P2P2(I)=0.07957747155D0*ALPMUL**3.5*ASQ**2*YY1
      YY0=DGASPD(FUNC,M1,M2,C0,D0,E0,F0,IND,1,1,0,ASQ,ALPHA1,ALPHA1)
      YY1=DGASPD(FUNC,M1,M2,C0,D0,E0,F0,IND,2,1,1,ASQ,ALPHA1,ALPHA1)
      YY2=DGASPD(FUNC,M1,M2,C0,D0,E0,F0,IND,2,1,0,ASQ,ALPHA1,ALPHA1)
```

5 PROGRAM LISTINGS

```
      YY3=DGASPD(FUNC,M1,M2,CO,DO,EO,FO,IND,3,2,0,ASQ,ALPHA1,ALPHA1)
      YY4=DGASPD(FUNC,M1,M2,CO,DO,EO,FO,IND,3,1,1,ASQ,ALPHA1,ALPHA1)
      U2S2P(I)=-0.09188814924D0*ALPHA1**6*ASQ*(YY0-0.5D0*ALPHA1**2*
     1 ASQ*YY1)
      U2P2P1(I)=0.3183098862D0*ALPHA1**7*ASQ*(YY2-0.125D0*ASQ*
     1 ALPHA1**2*(YY3+YY4))
      U2P2P2(I)=0.07957747155D0*ALPHA1**7*ASQ**2*YY1
  510 CONTINUE
  499 CONTINUE
      RETURN
      END



      SUBROUTINE DHQ4(M,IPARAM)
      IMPLICIT REAL*8 (A-H,O-Z)
      REAL*4 R,YY
      COMPLEX*16 A,LAMBDA,MULT,H,SFT(2),SHIFT,TEMP,TEMP1,TEMP2,ASIN,ACOS
     1 ,BSIN,BCOS,ASUM,AMUL,ZN
      COMMON/EORK/MULT(80),H(160,80),LAMBDA(80),INT(160)
      COMMON/COMP/A(80,80),BGAUSS,IER
      INTEGER IX*2
      EQUIVALENCE (R,IX)
      IER=0
      N=M
      IF(N.NE.1) GO TO 1
      LAMBDA(1)=A(1,1)
      A(1,1)=(1.D0,0.D0)
      RETURN
    1 ICOUNT=0
      N1=N-1
      NCAL=N
      SHIFT=(0.D0,0.D0)
      IF(N.NE.2) GO TO 4
    2 ASUM=A(1,1)+A(2,2)
      AMUL=A(1,1)*A(2,2)-A(1,2)*A(2,1)
      TEMP=(ASUM+CDSQRT(ASUM*ASUM-4.D0*AMUL))*0.5D0
      IF(DREAL(TEMP).NE.O.DO.OR.DIMAG(TEMP).NE.O.DO) GO TO 3
      LAMBDA(M)=SHIFT
      LAMBDA(M-1)=ASUM+SHIFT
      GO TO 37
    4 DO 15 K=1,N-2
      K1=K+1
      K2=K+2
      ABIG=0.D0
      INT(K)=K1
      DO 5 I=K1,N
      ABSSQ=DREAL(A(I,K))**2+DIMAG(A(I,K))**2
      IF(ABSSQ.LE.ABIG) GO TO 5
      INT(K)=I
      ABIG=ABSSQ
    5 CONTINUE
      INTER=INT(K)
      IF(ABIG.EQ.0.D0) GO TO 15
      IF(INTER.EQ.K1) GO TO 8
      DO 6 J=K,N
      TEMP=A(K1,J)
      A(K1,J)=A(INTER,J)
      A(INTER,J)=TEMP
```

```
   6 CONTINUE
     DO 7 I=1,N
     TEMP=A(I,K1)
     A(I,K1)=A(I,INTER)
     A(I,INTER)=TEMP
   7 CONTINUE
   8 ZN=1.D0/A(K1,K)
     DO 9 I=K2,N
     MULT(I)=A(I,K)*ZN
     A(I,K)=MULT(I)
   9 CONTINUE
     DO 11 J=K2,N
     DO 10 I=1,K1
     A(I,K1)=A(I,K1)+A(I,J)*MULT(J)
  10 CONTINUE
  11 CONTINUE
     DO 13 J=K2,N
     DO 12 I=K2,N
     A(I,K1)=A(I,K1)+A(I,J)*MULT(J)
  12 CONTINUE
  13 CONTINUE
     DO 60 I=K2,N
     A(I,K1)=A(I,K1)-MULT(I)*A(K1,K1)
  60 CONTINUE
     DO 14 J=K2,N
     DO 14 I=K2,N
     A(I,J)=A(I,J)-MULT(I)*A(K1,J)
  14 CONTINUE
  15 CONTINUE
     EPS=0.D0
     DO 16 J=1,N
     EPS=EPS+CDABS(A(1,J))
  16 CONTINUE
     DO 18 I=2,N
     SUM=0.D0
     I1=I-1
     DO 17 J=I1,N
     SUM=SUM+CDABS(A(I,J))
  17 CONTINUE
     IF(SUM.GT.EPS) EPS=SUM
  18 CONTINUE
C        E=SQRT(FLOAT(N))
     YY=FLOAT(N)
     R=YY
     IX=(IX+16640)/2
     R=(YY/(YY/R+R)*0.5+(YY/R+R)*0.5)*0.5
     R=(YY/(YY/R+R)*0.5+(YY/R+R)*0.5)*0.5
     E=(YY/(YY/R+R)*0.5+(YY/R+R)*0.5)*0.5
     IF(EPS.EQ.0.D0) EPS=1.D-12
     EPS=E*EPS*1.D-12
     IF(IPARAM.EQ.0) GO TO 20
     DO 19 J=1,N
     DO 19 I=1,N
     H(N+I,J)=A(I,J)
  19 CONTINUE
  20 IF(N.NE.1) GO TO 21
     LAMBDA(M)=A(1,1)+SHIFT
     GO TO 37
  21 IF(N.EQ.2) GO TO 2
```

```
   22 IF(DREAL(A(N,N)).EQ.O.DO.AND.DIMAG(A(N,N)).EQ.O.DO) GO TO 23
      TEMP=A(N,N1)/A(N,N)
      IF(DABS(DREAL(TEMP))+DABS(DIMAG(TEMP))-1.D-9) 24,24,23
   23 IF(DABS(DREAL(A(N,N1)))+DABS(DIMAG(A(N,N1))).GE.EPS) GO TO 25
   24 LAMBDA(M-N+1)=A(N,N)+SHIFT
      ICOUNT=O
      N=N-1
      N1=N-1
      GO TO 21
   25 ASUM=A(N1,N1)+A(N,N)
      AMUL=A(N1,N1)*A(N,N)-A(N1,N)*A(N,N1)
      TEMP=ASUM*ASUM-4.DO*AMUL
      SFT(1)=(ASUM+CDSQRT(TEMP))*0.5DO
      SFT(2)=AMUL/SFT(1)
      IF(DREAL(SFT(1)).EQ.O.DO.AND.DIMAG(SFT(1)).EQ.O.DO) SFT(2)=ASUM
      TEMP1=SFT(1)-A(N,N)
      TEMP2=SFT(2)-A(N,N)
      INDEX=1
      IF(CDABS(TEMP1).GE.CDABS(TEMP2)) INDEX=2
      IF(CDABS(A(N1,N-2)).GE.EPS) GO TO 30
      LAMBDA(M-N+1)=SFT(1)+SHIFT
      LAMBDA(M-N+2)=SFT(2)+SHIFT
      ICOUNT=O
      N=N-2
      N1=N-1
      GO TO 20
   30 SHIFT=SHIFT+SFT(INDEX)
      DO 31 I=1,N
      A(I,I)=A(I,I)-SFT(INDEX)
   31 CONTINUE
C
      IF(ICOUNT.LE.10) GO TO 32
      NCAL=M-N
      IER=1
      GO TO 37
   32 DO 36 K=1,N1
      K1=K+1
      K2=K+2
      TEMP1=A(K,K)
      TEMP2=A(K1,K)
      ROOT=DSQRT(DREAL(TEMP1)**2+DIMAG(TEMP1)**2+DREAL(TEMP2)**2
     1 +DIMAG(TEMP2)**2)
      IF(ROOT.EQ.O.DO) GO TO 36
      ACOS=TEMP1/ROOT
      ASIN=TEMP2/ROOT
      BCOS=DCONJG(ACOS)
      BSIN=DCONJG(ASIN)
      INDEX=MAXO(K-1,1)
      DO 33 J=INDEX,N
      TEMP=A(K,J)
      A(K,J)=BCOS*A(K,J)+BSIN*A(K1,J)
      A(K1,J)=-ASIN*TEMP+ACOS*A(K1,J)
   33 CONTINUE
      DO 34 I=1,K
      TEMP=A(I,K)
      A(I,K)=ACOS*A(I,K)+ASIN*A(I,K1)
      A(I,K1)=-BSIN*TEMP+BCOS*A(I,K1)
   34 CONTINUE
      INDEX=MINO(K2,N)
```

```
      DO 35 I=K1,INDEX
      A(I,K)=ASIN*A(I,K1)
      A(I,K1)=BCOS*A(I,K1)
   35 CONTINUE
   36 CONTINUE
      ICOUNT=ICOUNT+1
      GO TO 22
    3 LAMBDA(M)=TEMP+SHIFT
      LAMBDA(M-1)=AMUL/TEMP+SHIFT
   37 IF(NCAL.EQ.0) GO TO 59
      IF(IPARAM.EQ.0) RETURN
      N=M
      N1=N-1
      IF(N.NE.2) GO TO 38
      EPS=DMAX1(CDABS(LAMBDA(1)),CDABS(LAMBDA(2)))*1.D-8
      IF(EPS.EQ.0.D0) EPS=1.D-12
      H(3,1)=A(1,1)
      H(3,2)=A(1,2)
      H(4,1)=A(2,1)
      H(4,2)=A(2,2)
   38 DO 57 K=1,NCAL
      DO 40 J=1,N
      DO 39 I=1,N
      H(I,J)=H(N+I,J)
   39 CONTINUE
   40 CONTINUE
      DO 61 I=1,N
      H(I,I)=H(I,I)-LAMDDH(K)
   61 CONTINUE
      DO 44 I=1,N1
      NI=N+I
      I1=I+1
      MULT(I)=(0.D0,0.D0)
      INT(NI)=0
      IF(CDABS(H(I1,I)).LE.CDABS(H(I,I))) GO TO 42
      INT(NI)=1
      DO 41 J=I,N
      TEMP=H(I,J)
      H(I,J)=H(I1,J)
      H(I1,J)=TEMP
   41 CONTINUE
   42 IF(DREAL(H(I,I)).EQ.0.D0.AND.DIMAG(H(I,I)).EQ.0.D0) GO TO 44
      MULT(I)=-H(I1,I)/H(I,I)
      DO 43 J=I1,N
      H(I1,J)=H(I1,J)+MULT(I)*H(I,J)
   43 CONTINUE
   44 CONTINUE
C           START INVERSE ITERATION
      DO 45 I=1,N
      IF(DREAL(H(I,I)).EQ.0.D0.AND.DIMAG(H(I,I)).EQ.0.D0)
     1 H(I,I)=DCMPLX(EPS,0.D0)
      A(I,K)=(1.D0,0.D0)
   45 CONTINUE
      IIII=0
   46 CONTINUE
      IIII=IIII+1
      A(N,K)=A(N,K)/H(N,N)
      DO 48 I=1,N1
      NI=N-I
```

```
      DO 47 J=NI,N1
      J1=J+1
      A(NI,K)=A(NI,K)-H(NI,J1)*A(J1,K)
   47 CONTINUE
      A(NI,K)=A(NI,K)/H(NI,NI)
   48 CONTINUE
      IF(IIII.EQ.3) GO TO 52
      BIG=0.D0
      DO 49 I=1,N
      SUM=DABS(DREAL(A(I,K)))+DABS(DIMAG(A(I,K)))
      IF(BIG.LT.SUM) BIG=SUM
   49 CONTINUE
      ZN=1.D0/BIG
      DO 50 I=1,N
      A(I,K)=A(I,K)*ZN
   50 CONTINUE
      DO 51 I=1,N1
      NI=N+I
      I1=I+1
      IF(INT(NI).EQ.0) GO TO 51
      TEMP=A(I,K)
      A(I,K)=A(I1,K)
      A(I1,K)=TEMP
   51 A(I1,K)=A(I1,K)+MULT(I)*A(I,K)
      GO TO 46
C           TRANSFORM EIGENVECTORS
   52 IF(N.EQ.2) GO TO 55
      N2=N-2
      DO 54 J=1,N2
      NP=N-J+1
      NM=N-J-1
      NM1=NM+1
      DO 53 I=NP,N
      NI=N+I
      A(I,K)=A(I,K)+H(NI,NM)*A(NM1,K)
   53 CONTINUE
      INDEX=INT(NM)
      TEMP=A(NM1,K)
      A(NM1,K)=A(INDEX,K)
      A(INDEX,K)=TEMP
   54 CONTINUE
   55 CONTINUE
C           NORMALIZE
      SUM=0.D0
      DO 70 I=1,N
   70 SUM=SUM+CDABS(A(I,K))**2
      DO 71 I=1,N
   71 A(I,K)=A(I,K)/DSQRT(SUM)
   57 CONTINUE
      IF(IER.EQ.0) RETURN
      NCAL1=NCAL+1
      DO 58 K=NCAL1,N
      DO 58 I=1,N
      A(I,K)=(0.D0,0.D0)
   58 CONTINUE
      RETURN
   59 IER=2
      RETURN
      END
```

```
      SUBROUTINE EIGOUT(M,K)
      IMPLICIT REAL*8 (A-H,O-Z)
      COMMON/GAB/XXX(240),EPSILN(80),YYY(1680)
      WRITE(6,10) (EPSILN(I),I=M,K)
   10 FORMAT(//,15H EIGENVALUES   ,5(7X,F9.5,7X),//)
      RETURN
      END



      SUBROUTINE SCFOUT(OP)
      IMPLICIT REAL*8 (A-H,O-Z)
      COMPLEX*16 A
      COMMON/INFO/NATOMS,CHARGE,MULTIP,AN(35),C(35,3),N
      COMMON/INFO1/CZ(35),U(80),ULIM(35),LLIM(35),NELECS,OCCA,OCCB
      COMMON/ORB/ORB(9)
      COMMON/PERTBL/EL(18),NAME(20)
      COMMON/COMP/A(80,80),BGAUSS,IER
      COMMON/GAB/XXX(2000)
      INTEGER OP,AN,ANII,CZ,U,ORB,ULIM,EL,CHARGE,OCCA,OCCB
      DO 120 M=1,N,5
      K=M+4
      IF(K.LE.N) GO TO 30
   20 K=N
   30 CONTINUE
      WRITE(6,100)
      IF(OP.EQ.1) GO TO 40
      GO TO 50
   40 CALL EIGOUT(M,K)
   50 CONTINUE
      WRITE(6,60) (I,I=M,K)
   60 FORMAT(1H ,13X,5(9X,I2,12X),//)
      DO 110 I=1,N
      II=U(I)
      ANII=AN(II)
      L=I-LLIM(II)+1
   70 WRITE(6,80) I,II,EL(ANII),ORB(L),(A(I,J),J=M,K)
   80 FORMAT(1H ,I3,I3,2A4,5('(',F9.5,',',F9.5,')',2X))
      IF(I.EQ.ULIM(II)) GO TO 90
      GO TO 110
   90 WRITE(6,100)
  100 FORMAT(1X)
  110 CONTINUE
  120 CONTINUE
      WRITE(6,100)
      WRITE(6,100)
      RETURN
      END



      SUBROUTINE MATOUT(N,MATOP)
      IMPLICIT REAL*8 (A-H,O-Z)
      COMMON/ARRAYS/A(80,80,3)
      DO 80 M=1,N,11
      K=M+10
      IF(K.LE.N) GO TO 30
   20 K=N
   30 CONTINUE
```

```
      WRITE(6,40) (J,J=M,K)
   40 FORMAT(//,7X,11(4X,I2,3X),//)
      DO 60 I=1,N
      WRITE(6,50) I,(A(I,J,MATOP),J=M,K)
   50 FORMAT(1X,I2,4X,50(F9.4))
   60 CONTINUE
      WRITE(6,70)
   70 FORMAT(//)
   80 CONTINUE
      RETURN
      END



      SUBROUTINE YPRINT(M,N)
      IMPLICIT REAL*8 (A-H,O-Z)
      COMMON/GAUSTR/V1S2P(100),V2S2P(100),V2P2S(100),V2P2P1(100),V2P2P2(
     1 100),T1S2P(100),T2S2P(100),T2P2P1(100),T2P2P2(100),U2S2P(100),U2P
     2 2P1(100),U2P2P2(100),INDX1(35,35),INDX2(35,35),INVJ1I(100),INVJ1J
     3 (100),INVJ2I(100),INVJ2J(100),INVT1I(100),INVT1J(100),INVT2I(100)
     4 ,INVT2J(100),DXX(35,35),Y1(80,3,3),Y2(80,3,3)
      WRITE(6,1)
    1 FORMAT(1H0,8X,'Y1',/)
      J=0
    5 CONTINUE
      N1=1+10*J
      N2=N1+9
      IF(N.LE.N2) N2=N
      WRITE(6,2) (I,I=N1,N2)
    2 FORMAT(1H0,14X,10(I2,8X))
      DO 3 L=1,3
      WRITE(6,4)(Y1(I,L,M),I=N1,N2)
    4 FORMAT(1H ,10X,10F10.6)
    3 CONTINUE
      IF(N2.EQ.N) GO TO 6
      J=J+1
      GO TO 5
    6 CONTINUE
      WRITE(6,7)
    7 FORMAT(1H0,8X,'Y2',/)
      J=0
   12 CONTINUE
      N1=1+10*J
      N2=N1+9
      IF(N.LE.N2) N2=N
      WRITE(6,8) (I,I=N1,N2)
    8 FORMAT(1H0,14X,10(I2,8X))
      DO 9 L=1,3
      WRITE(6,10) (Y2(I,L,M),I=N1,N2)
   10 FORMAT(1H ,10X,10F10.6)
    9 CONTINUE
      IF(N2.EQ.N) GO TO 11
      J=J+1
      GO TO 12
   11 CONTINUE
      RETURN
      END
```

```
      SUBROUTINE CHSFT(LLL)
      IMPLICIT REAL*8 (A-H,O-Z)
      COMMON/ARRAYS/A(80,80),B(80,80),D(80,80)
      COMMON/GAB/XXX(400),G(35,35),Q(80),YYY(80),ENERGY,XXY(214)
      COMMON/INFO/NATOMS,CHARGE,MULTIP,AN(35),C(35,3),N
      COMMON/INFO1/CZ(35),U(80),ULIM(35),LLIM(35),NELECS,OCCA,OCCB
      COMMON/PERTBL/EL(18),NAME(20)
      COMMON/OPTION/OPTION,OPNCLO,HUCKEL,CNDO,INDO,CLOSED,OPEN,CONTR,MON
     1O,NEIG
      COMMON/COMP/BCOMPL(80,80),BGAUSS,IER
      COMMON/EORK/WORK1(80),WORK2(12800),LAMBDA(80),IWORK(160)
      COMMON/IMAG/AIMAGE(80,80),BIMAGE(80,80),XIMAG(80)
      COMMON/GAUSTR/V1S2P(100),V2S2P(100),V2P2S(100),V2P2P1(100),V2P2P2(
     1 100),T1S2P(100),T2S2P(100),T2P2P1(100),T2P2P2(100),U2S2P(100),U2P
     2 2P1(100),U2P2P2(100),INDX1(35,35),INDX2(35,35),INVJ1I(100),INVJ1J
     3 (100),INVJ2I(100),INVJ2J(100),INVT1I(100),INVT1J(100),INVT2I(100)
     4 ,INVT2J(100),DXX(35,35),Y1(80,3,3),Y2(80,3,3)
      DIMENSION DIAMAG(20),PARAMG(20),CTOTAL(20)
      DIMENSION TENSOR(3,3,20),EIGVAL(3),EIGVEC(3,3)
      COMPLEX*16 BCOMPL
      COMPLEX*16 WORK1,WORK2,LAMBDA
      INTEGER OPTION,OPNCLO,HUCKEL,CNDO,INDO,CLOSED,OPEN,CONTR,MONO,NEIG
      INTEGER CHARGE,AN,U,ULIM,EL,OCCA,OCCB,UL,CZ,ANI
      IF(IER.EQ.0) GO TO 1
      WRITE(6,2) IER
    2 FORMAT(1H0,5X,' THE RESULT IS NOT OBTAINED.',5X,'IER =',I2)
      RETURN
    1 CONTINUE
      WRITE(6,10)
   10 FORMAT(1H0,15H DENSITY MATRIX)
      CALL SCFOUT(0)
      IF(LLL.NE.1) GO TO 1000
      DO 40 I=1,NATOMS
      TCHG=0.D0
      LL=LLIM(I)
      UL=ULIM(I)
      DO 50 J=LL,UL
   50 TCHG=TCHG+B(J,J)
      XXY(I)=TCHG
   40 CONTINUE
      IF(BGAUSS.GT.0.01D0) GO TO 1000
      LLL=4
      GO TO 1001
 1000 CONTINUE
      NCARBN=0
      ACONST=1.251695D7/BGAUSS
      DO 600 II=1,NATOMS
      IF(AN(II).EQ. 5) GO TO 9600
      IF(AN(II).EQ. 6) GO TO 9600
      IF(AN(II).EQ. 7) GO TO 9600
      IF(AN(II).EQ. 8) GO TO 9600
      IF(AN(II).EQ. 9) GO TO 9600
      GO TO 600
 9600 CONTINUE
      NCARBN=NCARBN+1
      Y1T1=0.D0
      Y1T2=0.D0
      Y1T3=0.D0
      Y2T1=0.D0
      Y2T2=0.D0
```

```
      Y2T3=0.D0
      I0=LLIM(II)
      WRITE(6,610) II
  610 FORMAT(1H0,5X,'CENTERED ON ATOM-',I2,/)
      IF(AN(II).EQ.5)CONST=0.512D0*ACONST
      IF(AN(II).EQ.6)CONST=ACONST
      IF(AN(II).EQ.7)CONST=1.728D0*ACONST
      IF(AN(II).EQ.8)CONST=2.744D0*ACONST
      IF(AN(II).EQ.9)CONST=4.096D0*ACONST
      DO 680 K=1,3
      DO 680 J=1,3
      DO 680 I=1,N
      Y1(I,J,K)=0.D0
      Y2(I,J,K)=0.D0
  680 CONTINUE
      DO 620 JJ=1,NATOMS
      IF(CONTR.EQ.MONO)GO TO 9999
      J0=LLIM(JJ)
      C1=C(JJ,1)-C(II,1)
      C2=C(JJ,2)-C(II,2)
      C3=C(JJ,3)-C(II,3)
      INDEX=INDX2(II,JJ)
      IF(AN(JJ).GT.2) GO TO 640
      Y2(J0,2,1)= CONST*C3*T1S2P(INDEX)*BIMAGE(J0,I0+2)
      Y2(J0,3,1)=-CONST*C2*T1S2P(INDEX)*BIMAGE(J0,I0+3)
      Y2(J0,1,2)=-CONST*C3*T1S2P(INDEX)*BIMAGE(J0,I0+1)
      Y2(J0,3,2)= CONST*C1*T1S2P(INDEX)*BIMAGE(J0,I0+3)
      Y2(J0,1,3)= CONST*C2*T1S2P(INDEX)*BIMAGE(J0,I0+1)
      Y2(J0,2,3)=-CONST*C1*T1S2P(INDEX)*BIMAGE(J0,I0+2)
      Y2T1=Y2T1+Y2(J0,2,1)+Y2(J0,3,1)
      Y2T2=Y2T2+Y2(J0,1,2)+Y2(J0,3,2)
      Y2T3=Y2T3+Y2(J0,1,3)+Y2(J0,2,3)
      GO TO 620
  640 CONTINUE
      IF(II.EQ.JJ) GO TO 630
      Y2(J0,2,1)= CONST*C3*T2S2P(INDEX)*BIMAGE(J0,I0+2)
      Y2(J0,3,1)=-CONST*C2*T2S2P(INDEX)*BIMAGE(J0,I0+3)
      Y2(J0,1,2)=-CONST*C3*T2S2P(INDEX)*BIMAGE(J0,I0+1)
      Y2(J0,3,2)= CONST*C1*T2S2P(INDEX)*BIMAGE(J0,I0+3)
      Y2(J0,1,3)= CONST*C2*T2S2P(INDEX)*BIMAGE(J0,I0+1)
      Y2(J0,2,3)=-CONST*C1*T2S2P(INDEX)*BIMAGE(J0,I0+2)
      Y2(J0+1,2,1)= CONST*C1*C3*T2P2P1(INDEX)*BIMAGE(J0+1,I0+2)
      Y2(J0+1,3,1)=-CONST*C1*C2*T2P2P1(INDEX)*BIMAGE(J0+1,I0+3)
      Y2(J0+2,1,2)=-CONST*C2*C3*T2P2P1(INDEX)*BIMAGE(J0+2,I0+1)
      Y2(J0+2,3,2)= CONST*C1*C2*T2P2P1(INDEX)*BIMAGE(J0+2,I0+3)
      Y2(J0+3,1,3)= CONST*C2*C3*T2P2P1(INDEX)*BIMAGE(J0+3,I0+1)
      Y2(J0+3,2,3)=-CONST*C1*C3*T2P2P1(INDEX)*BIMAGE(J0+3,I0+2)
      Y2(J0+3,2,1)= CONST*(C3**2*T2P2P1(INDEX)+T2P2P2(INDEX))*BIMAGE
     1 (J0+3,I0+2)
      Y2(J0+2,3,1)=-CONST*(C2**2*T2P2P1(INDEX)+T2P2P2(INDEX))*BIMAGE
     1 (J0+2,I0+3)
      Y2(J0+3,1,2)=-CONST*(C3**2*T2P2P1(INDEX)+T2P2P2(INDEX))*BIMAGE
     1 (J0+3,I0+1)
      Y2(J0+1,3,2)= CONST*(C1**2*T2P2P1(INDEX)+T2P2P2(INDEX))*BIMAGE
     1 (J0+1,I0+3)
      Y2(J0+2,1,3)= CONST*(C2**2*T2P2P1(INDEX)+T2P2P2(INDEX))*BIMAGE
     1 (J0+2,I0+1)
      Y2(J0+1,2,3)=-CONST*(C1**2*T2P2P1(INDEX)+T2P2P2(INDEX))*BIMAGE
     1 (J0+1,I0+2)
      Y1(J0,2,1)= CONST*C3*U2S2P(INDEX)*BIMAGE(J0,I0+2)
```

```
      Y1(J0,3,1)=-CONST*C2*U2S2P(INDEX)*BIMAGE(J0,I0+3)
      Y1(J0,1,2)=-CONST*C3*U2S2P(INDEX)*BIMAGE(J0,I0+1)
      Y1(J0,3,2)= CONST*C1*U2S2P(INDEX)*BIMAGE(J0,I0+3)
      Y1(J0,1,3)= CONST*C2*U2S2P(INDEX)*BIMAGE(J0,I0+1)
      Y1(J0,2,3)=-CONST*C1*U2S2P(INDEX)*BIMAGE(J0,I0+2)
      Y1(J0+1,2,1)= CONST*C1*C3*U2P2P1(INDEX)*BIMAGE(J0+1,I0+2)
      Y1(J0+1,3,1)=-CONST*C1*C2*U2P2P1(INDEX)*BIMAGE(J0+1,I0+3)
      Y1(J0+2,1,2)=-CONST*C2*C3*U2P2P1(INDEX)*BIMAGE(J0+2,I0+1)
      Y1(J0+2,3,2)= CONST*C1*C2*U2P2P1(INDEX)*BIMAGE(J0+2,I0+3)
      Y1(J0+3,1,3)= CONST*C2*C3*U2P2P1(INDEX)*BIMAGE(J0+3,I0+1)
      Y1(J0+3,2,3)=-CONST*C1*C3*U2P2P1(INDEX)*BIMAGE(J0+3,I0+2)
      Y1(J0+3,2,1)= CONST*((C2**2+C3**2)*U2P2P1(INDEX)+U2P2P2(INDEX))*
     1 BIMAGE(J0+3,J0+2)
      Y1(J0+3,1,2)=-CONST*((C1**2+C3**2)*U2P2P1(INDEX)+U2P2P2(INDEX))*
     1 BIMAGE(J0+3,J0+1)
      Y1(J0+2,1,3)= CONST*((C1**2+C2**2)*U2P2P1(INDEX)+U2P2P2(INDEX))*
     1 BIMAGE(J0+2,J0+1)
      Y1(J0+2,1,1)=Y1(J0+1,2,1)
      Y1(J0+2,3,1)=Y1(J0+3,2,1)
      Y1(J0+3,1,1)=Y1(J0+1,3,1)
      Y1(J0+1,2,2)=Y1(J0+2,1,2)
      Y1(J0+1,3,2)=Y1(J0+3,1,2)
      Y1(J0+3,2,2)=Y1(J0+2,3,2)
      Y1(J0+1,2,3)=Y1(J0+2,1,3)
      Y1(J0+1,3,3)=Y1(J0+3,1,3)
      Y1(J0+2,3,3)=Y1(J0+3,2,3)
      Y1T1=Y1T1+Y1(J0,2,1)+Y1(J0,3,1)
      Y1T2=Y1T2+Y1(J0,1,2)+Y1(J0,3,2)
      Y1T3=Y1T3+Y1(J0,1,3)+Y1(J0,2,3)
      DO 690 K1=1,4
      K=J0+K1-1
      DO 690 L=1,3
      Y2T1=Y2T1+Y2(K,L,1)
      Y2T2=Y2T2+Y2(K,L,2)
      Y2T3=Y2T3+Y2(K,L,3)
      Y1T1=Y1T1+Y1(K,L,1)
      Y1T2=Y1T2+Y1(K,L,2)
      Y1T3=Y1T3+Y1(K,L,3)
  690 CONTINUE
      GO TO 620
  630 CONTINUE
 9999 CONTINUE
      Y1(I0+3,2,1)=1.430339D0*CONST*BIMAGE(I0+3,I0+2)
      Y1(I0+2,3,1)=Y1(I0+3,2,1)
      Y1(I0+1,3,2)=1.430339D0*CONST*BIMAGE(I0+1,I0+3)
      Y1(I0+3,1,2)=Y1(I0+1,3,2)
      Y1(I0+2,1,3)=1.430339D0*CONST*BIMAGE(I0+2,I0+1)
      Y1(I0+1,2,3)=Y1(I0+2,1,3)
      Y0T1=2.D0*Y1(I0+3,2,1)
      Y0T2=2.D0*Y1(I0+1,3,2)
      Y0T3=2.D0*Y1(I0+2,1,3)
  620 CONTINUE
      Y2T1=2.D0*Y2T1
      Y2T2=2.D0*Y2T2
      Y2T3=2.D0*Y2T3
C
      WRITE(6,650) LLL
  650 FORMAT(1H0,10X,'*** X COMPONENT ***    (LLL=',I2,')',/)
      CALL YPRINT(1,N)
```

```
      WRITE(6,655) YOT1,Y1T1,Y2T1
  655 FORMAT(1H0,10X,'Y0=',F10.2,'Y1=',F10.2,'Y2=',F10.2)
      WRITE(6,660) LLL
  660 FORMAT(1H0,10X,'***   Y  COMPONENT   ***        (LLL=',I2,')',/)
      CALL YPRINT(2,N)
      WRITE(6,655) YOT2,Y1T2,Y2T2
      WRITE(6,670) LLL
  670 FORMAT(1H0,10X,'***  Z COMPONENT  ***    (LLL=',I2,')',/)
      CALL YPRINT(3,N)
      WRITE(6,655) YOT3,Y1T3,Y2T3
      TENSOR(1,LLL,NCARBN)=YOT1+Y1T1+Y2T1
      TENSOR(2,LLL,NCARBN)=YOT2+Y1T2+Y2T2
      TENSOR(3,LLL,NCARBN)=YOT3+Y1T3+Y2T3
  600 CONTINUE
      IF(LLL.LE.2) RETURN
 1001 CONTINUE
      DO 1002 I=1,NATOMS
      ANI=AN(I)
      WRITE(6,1003) I,EL(ANI),XXY(I)
 1003 FORMAT(1H0,I3,A4,8X,F7.4)
 1002 CONTINUE
      IF(BGAUSS.LT.0.01D0) RETURN
      WRITE(6,80)
   80 FORMAT(1H1,'NUCLEAR SHIELDING (USING THE FINITE PERTURBATIO
     1N METHOD',/)
      WRITE(6,90) (NAME(I),I=1,19)
   90 FORMAT(1H ,19A4)
      WRITE(6,100)
  100 FORMAT(1H ,4X,'PARAMAGNETIC SHIELDING TENSOR '/)
      DO 200 I=1,NCARBN
      WRITE(6,210) I
  210 FORMAT(1H0,10X,'ATOM-' I2,/)
      WRITE(6,220)
  220 FORMAT(1H ,14X,'X',9X,'Y',9X,'Z',/)
      WRITE(6,230)(TENSOR(J,1,I),J=1,3)
  230 FORMAT(1H ,7X,'X',3(2X,F8.2))
      WRITE(6,240)(TENSOR(J,2,I),J=1,3)
  240 FORMAT(1H ,7X,'Y',3(2X,F8.2))
      WRITE(6,250)(TENSOR(J,3,I),J=1,3)
  250 FORMAT(1H ,7X,'Z',3(2X,F8.2))
      PARAMG(I)=0.D0
      DO 260 J=1,3
  260 PARAMG(I)=PARAMG(I)+TENSOR(J,J,I)
      PARAMG(I)=PARAMG(I)/3.D0
      DO 270 J=1,3
      DO 270 K=1,3
      BCOMPL(J,K)=DCMPLX(TENSOR(K,J,I),0.D0)
  270 CONTINUE
      CALL DHQ4(3,1)
      DO 280 J=1,3
      IF(DIMAG(LAMBDA(J)).LT.0.001D0) GO TO 280
      WRITE(6,290)
  290 FORMAT(1H ,'EIGENVALUE CONTAINS AN IMAGINARY PART.',/)
  280 CONTINUE
      WRITE(6,300)
  300 FORMAT(1H0,10X,'EIGENVALUE AND EIGENVECTOR',/)
      DO 310 J=1,3
      EIGVAL(J)=DREAL(LAMBDA(J))
      DO 310 K=1,3
```

```
      EIGVEC(J,K)=DREAL(BCOMPL(J,K))
310 CONTINUE
      WRITE(6,320)(EIGVAL(J),J=1,3)
320 FORMAT(1H ,15X,3F10.2,/)
      WRITE(6,330)(EIGVEC(1,K),K=1,3)
      WRITE(6,331)(EIGVEC(2,K),K=1,3)
      WRITE(6,332)(EIGVEC(3,K),K=1,3)
330 FORMAT(1H ,12X,'X     ',3F10.5)
331 FORMAT(1H ,12X,'Y     ',3F10.5)
332 FORMAT(1H ,12X,'Z     ',3F10.5)
      WRITE(6,340)
340 FORMAT(1H0,//)
200 CONTINUE
      WRITE(6,140)
140 FORMAT(1H0,4X,'ISOTROPIC NUCLEAR SHIELDING')
      WRITE(6,150)
150 FORMAT(1H0,7X,'I',3X,'DENSITY',4X,'DIAMAG',4X,'PARAMAG',5X,'TOTAL
     1 ',/)
      NCARBN=0
      DO 160 I=1,NATOMS
      IF(AN(I).EQ.1) GO TO 160
      NCARBN=NCARBN+1
      IF(AN(I).EQ.5)DIAMAG(NCARBN)=4.45D0*(2.6D0-0.35D0*(XXY(I)-3.0D0))*
     *XXY(I)
      IF(AN(I).EQ.6)DIAMAG(NCARBN)=4.45D0*(3.25D0-0.35D0*(XXY(I)-4.D0))*
     *XXY(I)
      IF(AN(I).EQ.7)DIAMAG(NCARBN)=4.45D0*(3.9D0-0.35D0*(XXY(I)-5.0D0))*
     *XXY(I)
      IF(AN(I).EQ.8)DIAMAG(NCARBN)=4.45D0*(4.55D0-0.35D0*(XXY(I)-6.0D0))
     **XXY(I)
      IF(AN(I).EQ.9)DIAMAG(NCARBN)=4.45D0*(5.20D0-0.35D0*(XXY(I)-7.0D0))
     **XXY(I)
      CTOTAL(NCARBN)=DIAMAG(NCARBN)+PARAMG(NCARBN)
      WRITE(6,170) NCARBN,XXY(I),DIAMAG(NCARBN),PARAMG(NCARBN),CTOTAL(NC
     1ARBN)
160 CONTINUE
170 FORMAT(1H ,6X,I2,3X,F7.4,4X,F7.2,5X,F7.2,5X,F7.2)
      RETURN
      END



      SUBROUTINE SCFCLO
      IMPLICIT REAL*8 (A-H,O-Z)
      COMMON/ARRAYS/A(80,80),B(80,80),D(80,80)
      COMMON/GAB/XXX(240),EIG(80),XXXX(80),G(35,35),Q(80),YYY(80),ENERGY
     1 ,XXY(214)
      COMMON/INFO/NATOMS,CHARGE,MULTIP,AN(35),C(35,3),N
      COMMON/INFO1/CZ(35),U(80),ULIM(35),LLIM(35),NELECS,OCCA,OCCB
      COMMON/OPTION/OPTION,OPNCLO,HUCKEL,CNDO,INDO,CLOSED,OPEN,CONTR,MON
      COMMON/IMAG/AIMAGE(80,80),BIMAGE(80,80),XIMAG(80)
      COMMON/COMP/BCOMPL(80,80),BGAUSS,IER
      COMMON/EORK/WORK1(80),WORK2(12800),LAMBDA(80),IWORK(160)
      COMMON/PERTBL/EL(18),NAME(20)
      COMMON/AA/AA(80,80)
      COMPLEX*16 BCOMPL,LAMBDA,WORK1,WORK2
      INTEGER OPTION,OPNCLO,HUCKEL,CNDO,INDO,CLOSED,OPEN,CONTR,MONO,NEIG
      INTEGER CHARGE,OCCA,OCCB,ULIM,U,AN,CZ,Z,EL
      INTEGER UL
```

```
      DIMENSION G1(18),F2(18)
      G1(3)=.092012D0
      G1(4)=.1407   D0
      G1(5)=.199265D0
      G1(6)=.267708D0
      G1(7)=.346029D0
      G1(8)=.43423  D0
      G1(9)=.532305D0
      F2(3)=.049865D0
      F2(4)=.089125D0
      F2(5)=.13041  D0
      F2(6)=.17372  D0
      F2(7)=.219055D0
      F2(8)=.266415D0
      F2(9)=.31580  D0
      Z=0
      IT=15
   10 CONTINUE
      IF(IER.NE.0) RETURN
      Z=Z+1
      ENERGY=0.D0
      DO 20 I=1,N
      A(I,I)=Q(I)
      DO 20 J=I,N
      A(J,I)=A(I,J)
   20 AIMAGE(J,I)=-AIMAGE(I,J)
      DO 30 I=1,N
      II=U(I)
      A(I,I)=A(I,I)-B(I,I)*G(II,II)*0.5D0
      DO 30 K=1,N
      JJ=U(K)
   30 A(I,I)=A(I,I)+B(K,K)*G(II,JJ)
      DO 35 I=1,N
   35 AIMAGE(I,I)=0.D0
      NM=N-1
      DO 40 I=1,NM
      II=U(I)
      LL=I+1
      DO 40 J=LL,N
      JJ=U(J)
      A(J,I)=A(J,I)-B(J,I)*G(II,JJ)*0.5D0
      AIMAGE(J,I)=AIMAGE(J,I)+BIMAGE(J,I)*G(II,JJ)*0.5D0
   40 CONTINUE
C     INDO MODIFICATION
      IF(OPTION.EQ.CNDO) GO TO 90
   50 DO 80 II=1,NATOMS
      K=AN(II)
      I=LLIM(II)
      IF(K.EQ.1) GO TO 80
   60 PAA=B(I,I)+B(I+1,I+1)+B(I+2,I+2)+B(I+3,I+3)
      A(I,I)=A(I,I)-(PAA-B(I,I))*G1(K)/6.D0
      DO 70 J=1,3
      A(I+J,I+J)=A(I+J,I+J)-B(I,I)*G1(K)/6.D0-(PAA-B(I,I))*7.D0*
     1 F2(K)/50.D0+B(I+J,I+J)*11.D0*F2(K)/50.D0
      A(I+J,I)=A(I+J,I)+B(I,I+J)*G1(K)/2.D0
      AIMAGE(I+J,I)=AIMAGE(I+J,I)-BIMAGE(I+J,I)*G1(K)/6.D0
   70 CONTINUE
      I1=I+1
      I2=I+2
```

```
            I3=I+3
            A(I2,I1)=A(I2,I1)+B(I2,I1)*11.D0*F2(K)/50.D0
            A(I3,I1)=A(I3,I1)+B(I3,I1)*11.D0*F2(K)/50.D0
            A(I3,I2)=A(I3,I2)+B(I3,I2)*11.D0*F2(K)/50.D0
            AIMAGE(I2,I1)=AIMAGE(I2,I1)-BIMAGE(I2,I1)*F2(K)/10.D0
            AIMAGE(I3,I1)=AIMAGE(I3,I1)-BIMAGE(I3,I1)*F2(K)/10.D0
            AIMAGE(I3,I2)=AIMAGE(I3,I2)-BIMAGE(I3,I2)*F2(K)/10.D0
         80 CONTINUE
         90 CONTINUE
            DO 100 I=1,N
        100 ENERGY=ENERGY+0.5D0*B(I,I)*(A(I,I)+Q(I))
            DO 105 I=1,NM
            LL=I+1
            DO 105 J=LL,N
        105 ENERGY=ENERGY+B(I,J)*(A(I,J)+A(J,I))
            WRITE(6,110) ENERGY
        110 FORMAT(//,10X,22H ELECTRONIC ENERGY      ,F16.10,2X,4HA.U.)
            IF(DABS(ENERGY-OLDENG).GE.0.00001D0) GO TO 150
        120 Z=16
        130 WRITE(6,140)
        140 FORMAT(5X,18H ENERGY SATISFIED /)
            GO TO 170
        150 CONTINUE
        160 OLDENG=ENERGY
        170 CONTINUE
C       SYMMETRIZE F FOR PRINTING (MATRIX ACOMPL)
        180 DO 190 I=1,N
            DO 190 J=I,N
            BCOMPL(J,I)=DCMPLX(A(J,I),AIMAGE(J,I))
            BCOMPL(I,J)=DCONJG(BCOMPL(J,I))
        190 CONTINUE
            IF(Z.LT.16) GO TO 210
            WRITE(6,200)
        200 FORMAT(1X,27H HARTREE-FOCK ENERGY MATRIX)
C           CALL SCFOUT(0)
        210 CONTINUE
            CALL DHQ4(N,1)
            DO 1003 I=1,N
            IF(DIMAG(LAMBDA(I)).GT.0.0001D0) IER=3
            IF(IER.NE.0) RETURN
       1003 EIG(I)=DREAL(LAMBDA(I))
            DO 1100 J=1,N
            AMINI=EIG(J)
            K1=J
            DO 1101 K=J,N
            IF(AMINI.LE.EIG(K)) GO TO 1101
            AMINI=EIG(K)
            K1=K
       1101 CONTINUE
            IF(K1.EQ.J) GO TO 1100
            EIG(K1)=EIG(J)
            EIG(J)=AMINI
            DO 1102 I=1,N
            WORK1(I)=BCOMPL(I,K1)
            BCOMPL(I,K1)=BCOMPL(I,J)
            BCOMPL(I,J)=WORK1(I)
       1102 CONTINUE
       1100 CONTINUE
            DO 1103 I=1,N
```

```
      DO 1103 J=1,N
      B(I,J)=DREAL(BCOMPL(I,J))
      BIMAGE(I,J)=DIMAG(BCOMPL(I,J))
 1103 CONTINUE
      IF(Z.LT.16) GO TO 240
  220 WRITE(6,230)
  230 FORMAT(1X,28HEIGENVALUES AND EIGENVECTORS)
C     CALL SCFOUT(1)
  240 CONTINUE
C     EIGENVECTORS (IN BCOMPL) ARE CONVERTED INTO DENSITY MATRIX (UN BCO
      DO 280 I=1,N
      DO 260 J=I,N
      XXX(J)=0.D0
      XIMAG(J)=0.D0
      DO 250 K=1,OCCA
      XXX(J)=XXX(J)+B(I,K)*B(J,K)+BIMAGE(I,K)*BIMAGE(J,K)
      XIMAG(J)=XIMAG(J)+B(I,K)*BIMAGE(J,K)-BIMAGE(I,K)*B(J,K)
  250 CONTINUE
      XXX(J)=2.D0*XXX(J)
      XIMAG(J)=2.D0*XIMAG(J)
  260 CONTINUE
      DO 270 J=I,N
      B(I,J)=XXX(J)
      BIMAGE(I,J)=XIMAG(J)
  270 CONTINUE
  280 CONTINUE
      DO 275 I=1,N
      DO 275 J=I,N
      B(J,I)=B(I,J)
      BIMAGE(J,I)=-BIMAGE(I,J)
  275 CONTINUE
      IF(Z.LE.IT) GO TO 10
      DO 290 I=1,N
      DO 290 J=I,N
      BCOMPL(I,J)=DCMPLX(B(I,J),BIMAGE(I,J))
  290 BCOMPL(J,I)=DCONJG(BCOMPL(I,J))
  300 CONTINUE
      RETURN
      END


      SUBROUTINE HUCKCL(LLL)
      IMPLICIT REAL*8 (A-H,O-Z)
      COMMON/ARRAYS/A(80,80),B(80,80),D(80,80)
      COMMON/GAB/XXX(240),EIG(80),XXXX(80),G(35,35),Q(80),YYY(80),ENERGY
     1 ,XXY(214)
      COMMON/INFO/NATOMS,CHARGE,MULTIP,AN(35),C(35,3),N
      COMMON/INFO1/CZ(35),U(80),ULIM(35),LLIM(35),NELECS,OCCA,OCCB
      COMMON/OPTION/OPTION,OPNCLO,HUCKEL,CNDO,INDO,CLOSED,OPEN,CONTR,MON
     10,NEIG
      COMMON/IMAG/AIMAGE(80,80),BIMAGE(80,80),XIMAG(80)
      COMMON/COMP/BCOMPL(80,80),BGAUSS,IER
      COMMON/EORK/WORK1(80),WORK2(12800),LAMBDA(80),IWORK(160)
      COMMON/PERTBL/EL(18),NAME(20)
      COMMON/GAUSTR/V1S2P(100),V2S2P(100),V2P2S(100),V2P2P1(100),V2P2P2(
     1 100),T1S2P(100),T2S2P(100),T2P2P1(100),T2P2P2(100),U2S2P(100),U2P
     2 2P1(100),U2P2P2(100),INDX1(35,35),INDX2(35,35),INVJ1I(100),INVJ1J
     3 (100),INVJ2I(100),INVJ2J(100),INVT1I(100),INVT1J(100),INVT2I(100)
```

```
4  ,INVT2J(100),DXX(35,35),Y1(80,3,3),Y2(80,3,3)
      COMPLEX*16 BCOMPL,LAMBDA,WORK1,WORK2
      INTEGER OPTION,OPNCLO,HUCKEL,CNDO,INDO,CLOSED,OPEN,CONTR,MONO,NEIG
      INTEGER CHARGE,OCCA,OCCB,UL,AN,CZ,U,ULIM,ANI,EL
      DIMENSION ENEG(18,3),BETA0(18)
      DIMENSION G1(18),F2(18)
      DIMENSION QQ(80)
      BCORR=1.D0
      IF(LLL.GT.1) GO TO 2002
      G1(3)=0.092012       D0
      G1(4)=0.1407         D0
      G1(5)=0.199265       D0
      G1(6)=0.267708       D0
      G1(7)=0.346029       D0
      G1(8)=0.43423        D0
      G1(9)=0.532305       D0
      F2(3)=0.049865       D0
      F2(4)=0.089125       D0
      F2(5)=0.13041        D0
      F2(6)=0.17372        D0
      F2(7)=0.219055       D0
      F2(8)=0.266415       D0
      F2(9)=0.31580        D0
      ENEG(1,1)=7.1761     D0
      ENEG(3,1)=3.1055     D0
      ENEG(3,2)=1.258      D0
      ENEG(4,1)=5.94557    D0
      ENEG(4,2)=2.563      D0
      ENEG(5,1)=9.59407    D0
      ENEG(5,2)=4.001      D0
      ENEG(6,1)=14.051     D0
      ENEG(6,2)=5.572      D0
      ENEG(7,1)=19.31637D0
      ENEG(7,2)=7.275      D0
      ENEG(8,1)=25.39017D0
      ENEG(8,2)=9.111      D0
      ENEG(9,1)=32.2724    D0
      ENEG(9,2)=11.08      D0
      ENEG(11,1)=2.804     D0
      ENEG(11,2)=1.302     D0
      ENEG(11,3)=0.150     D0
      ENEG(12,1)=5.1254    D0
      ENEG(12,2)=2.0516    D0
      ENEG(12,3)=0.16195D0
      ENEG(13,1)=7.7706D0
      ENEG(13,2)=2.9951D0
      ENEG(13,3)=0.22425D0
      ENEG(14,1)=10.0327D0
      ENEG(14,2)=4.1325    D0
      ENEG(14,3)=0.337     D0
      ENEG(15,1)=14.0327D0
      ENEG(15,2)=5.4638    D0
      ENEG(15,3)=0.500     D0
      ENEG(16,1)=17.6496D0
      ENEG(16,2)=6.989     D0
      ENEG(16,3)=0.71325D0
      ENEG(17,1)=21.5906D0
      ENEG(17,2)=8.7081    D0
      ENEG(17,3)=0.97695D0
```

```
      BETA0(1)= -13.        D0
      BETA0(3)= -9.         D0
      BETA0(4)= -13.        D0
      BETA0(5)= -17.        D0
      BETA0(6)= -15.        D0
      BETA0(7)= -25.        D0
      BETA0(8)= -31.        D0
      BETA0(9)= -39.        D0
      BETA0(11)=-7.7203 D0
      BETA0(12)=-9.4471 D0
      BETA0(13)=-11.3011D0
      BETA0(14)=-13.065 D0
      BETA0(15)=-15.070 D0
      BETA0(16)=-18.150 D0
      BETA0(17)=-22.330 D0
      NELECS=0
      DO 60 I=1,NATOMS
      NELECS=NELECS+CZ(I)
      LL=LLIM(I)
      UL=ULIM(I)
      ANI=AN(I)
      L=0
      DO 55 J=LL,UL
      L=L+1
      IF(L.EQ.1) GO TO 10
   20 IF(L.LT.5) GO TO 40
   30 A(J,J)=-ENEG(ANI,3)/27.21D0
      GO TO 50
   40 A(J,J)=-ENEG(ANI,2)/27.21D0
      GO TO 50
   10 A(J,J)=-ENEG(ANI,1)/27.21D0
   50 AIMAGE(J,J)=0.D0
      QQ(J)=A(J,J)
   55 CONTINUE
   60 CONTINUE
      DO 5 J=1,N
      DO 5 I=1,N
      AIMAGE(I,J)=0.D0
    5 CONTINUE
      NELECS=NELECS-CHARGE
      OCCA=NELECS/2
      DO 90 I=2,N
      K=U(I)
      L=AN(K)
      UL=I-1
      DO 90 J=1,UL
      KK=U(J)
      LL=AN(KK)
      IF((L.GT.9).OR.(LL.GT.9)) GO TO 70
      A(I,J)=A(I,J)*(BETA0(L)+BETA0(LL))/54.42D0
      IF((L.EQ.1).AND.(LL.EQ.6)) GO TO 80
      IF((L.EQ.6).AND.(LL.EQ.1)) GO TO 80
      GO TO 85
   80 A(I,J)=A(I,J)*BCORR
      GO TO 85
   70 A(I,J)=0.75D0*A(I,J)*(BETA0(L)+BETA0(LL))/54.42D0
   85 AIMAGE(I,J)=0.0
      A(J,I)=A(I,J)
      AIMAGE(J,I)=0.D0
```

```
      90 CONTINUE
         GO TO 2000
    2002 CONTINUE
         DO 2001 I=1,N
         A(I,I)=GQ(I)
         DO 2001 J=I,N
         A(J,I)=A(I,J)
         AIMAGE(I,J)=0.D0
         AIMAGE(J,I)=0.D0
    2001 CONTINUE
    2000 CONTINUE
         GO TO (96,97,98),LLL
      96 IIII=2
         JJJJ=3
         GO TO 99
      97 IIII=3
         JJJJ=1
         GO TO 99
      98 IIII=1
         JJJJ=2
      99 CONTINUE
         CONST=0.21273D-5*BGAUSS
         IF(CONTR.EQ.MONO)GO TO 9999
         DO 92 II=2,NATOMS
         IO=LLIM(II)
         IF(AN(II).GT.2) GO TO 93
         DO 94 JJ=1,II-1
         IF(AN(JJ).LT.3) GO TO 94
         JO=LLIM(JJ)
         INDEX=INDX1(II,JJ)
         RIIII=C(JJ,IIII)-C(II,IIII)
         RJJJJ=C(II,JJJJ)-C(JJ,JJJJ)
         AIMAGE(IO,JO+IIII)=CONST*RJJJJ*V1S2P(INDEX)
         AIMAGE(IO,JO+JJJJ)=CONST*RIIII*V1S2P(INDEX)
      94 CONTINUE
         GO TO 92
      93 CONTINUE
         DO 95 JJ=1,II-1
         JO=LLIM(JJ)
         INDEX=INDX1(II,JJ)
         RIIII=C(JJ,IIII)-C(II,IIII)
         RJJJJ=C(II,JJJJ)-C(JJ,JJJJ)
         IF(AN(JJ).GT.2) GO TO 101
         AIMAGE(IO+IIII,JO)=CONST*RJJJJ*V1S2P(INDEX)
         AIMAGE(IO+JJJJ,JO)=CONST*RIIII*V1S2P(INDEX)
         GO TO 95
     101 CONTINUE
         RLLL=C(II,LLL)-C(JJ,LLL)
         AIMAGE(IO+IIII,JO)=CONST*RJJJJ*V2P2S(INDEX)
         AIMAGE(IO+JJJJ,JO)=CONST*RIIII*V2P2S(INDEX)
         AIMAGE(IO,JO+IIII)=CONST*RJJJJ*V2S2P(INDEX)
         AIMAGE(IO,JO+JJJJ)=CONST*RIIII*V2S2P(INDEX)
         AIMAGE(IO+LLL,JO+IIII)=CONST*RLLL*RJJJJ*V2P2P2(INDEX)
         AIMAGE(IO+LLL,JO+JJJJ)=CONST*RLLL*RIIII*V2P2P2(INDEX)
         AIMAGE(IO+IIII,JO+LLL)=-AIMAGE(IO+LLL,JO+IIII)
         AIMAGE(IO+JJJJ,JO+LLL)=-AIMAGE(IO+LLL,JO+JJJJ)
         AIMAGE(IO+IIII,JO+JJJJ)= CONST*(V2P2P1(INDEX)+(RIIII**2+RJJJJ**2)*
       1 V2P2P2(INDEX))
         AIMAGE(IO+JJJJ,JO+IIII)=-AIMAGE(IO+IIII,JO+JJJJ)
```

```
   95 CONTINUE
   92 CONTINUE
      DO 102 I=2,N
      DO 102 J=1,I-1
      AIMAGE(J,I)=-AIMAGE(I,J)
  102 CONTINUE
 9999 CONTINUE
      DO 91 II=1,NATOMS
      IF(AN(II).LT.3) GO TO 91
      IO=LLIM(II)
      AIMAGE(IO+IIII,IO+JJJJ)=CONST
      AIMAGE(IO+JJJJ,IO+IIII)=-CONST
   91 CONTINUE
      DO 100 I=1,N
      Q(I)=A(I,I)
      DO 100 J=1,N
      BCOMPL(I,J)=DCMPLX(A(I,J),AIMAGE(I,J))
  100 CONTINUE
 2120 CONTINUE
      CALL DHQ4(N,1)
      IF(IER.NE.0) RETURN
      DO 1003 I=1,N
      IF(DIMAG(LAMBDA(I)).GT.0.0001D0) IER=3
      IF(IER.NE.0) RETURN
 1003 EIG(I)=DREAL(LAMBDA(I))
      DO 1100 J=1,N
      AMINI=EIG(J)
      K1=J
      DO 1101 K=J,N
      IF(AMINI.LE.EIG(K)) GO TO 1101
      AMINI=EIG(K)
      K1=K
 1101 CONTINUE
      IF(K1.EQ.J) GO TO 1100
      EIG(K1)=EIG(J)
      EIG(J)=AMINI
      DO 1102 I=1,N
      WORK1(I)=BCOMPL(I,K1)
      BCOMPL(I,K1)=BCOMPL(I,J)
      BCOMPL(I,J)=WORK1(I)
 1102 CONTINUE
 1100 CONTINUE
      EIGENVECTORS (IN BCOMPL) ARE CONVERTED INTO DENSITY MATRIX (IN BCO
      DO 105 I=1,N
      DO 105 J=1,N
      B(I,J)=DREAL(BCOMPL(I,J))
      BIMAGE(I,J)=DIMAG(BCOMPL(I,J))
  105 CONTINUE
      DO 140 I=1,N
      DO 120 J=I,N
      XXX(J)=0.D0
      XIMAG(J)=0.D0
      DO 110 K=1,OCCA
      XXX(J)=XXX(J)+2.D0*B(I,K)*B(J,K)+2.D0*BIMAGE(I,K)*BIMAGE(J,K)
  110 XIMAG(J)=XIMAG(J)+2.D0*B(I,K)*BIMAGE(J,K)-2.D0*B(J,K)*BIMAGE(I,K)
  120 CONTINUE
      DO 130 J=I,N
      B(I,J)=XXX(J)
  130 BIMAGE(I,J)=XIMAG(J)
```

```
      140 CONTINUE
          DO 130 I=1,N
          DO 150 J=I,N
          B(J,I)=B(I,J)
      150 BIMAGE(J,I)=-BIMAGE(I,J)
    C     ADD V(AB) TO HCORE---CNDO
          DO 170 I=1,N
          J=U(I)
          Q(I)=Q(I)+0.5D0*G(J,J)
          DO 160 K=1,NATOMS
      160 Q(I)=Q(I)-DFLOAT(CZ(K))*G(J,K)
      170 CONTINUE
          IF(OPTION.EQ.CNDO) GO TO 290
      180 DO 280 I=1,NATOMS
          K=AN(I)
          J=LLIM(I)
          IF((K.GT.1).AND.(K.LT.10)) GO TO 190
          GO TO 280
      190 IF(K.LE.3) GO TO 210
      200 Q(J)=Q(J)+(DFLOAT(CZ(I))-1.5D0)*G1(K)/6.D0
      210 IF(K.EQ.3) GO TO 220
      230 IF(K.EQ.4) GO TO 240
      250 TEMP=G1(K)/3.D0+(DFLOAT(CZ(I))-2.5D0)*2.D0*F2(K)/25.D0
          GO TO 260
      240 TEMP=G1(K)/4.D0
          GO TO 260
      220 TEMP=G1(K)/12.D0
      260 CONTINUE
          DO 270 L=1,3
      270 Q(J+L)=Q(J+L)+TEMP
      280 CONTINUE
      290 CONTINUE
          DO 300 I=1,N
      300 A(I,I)=Q(I)
          DO 310 I=1,N
          DO 310 J=1,N
      310 BCOMPL(I,J)=DCMPLX(A(I,J),AIMAGE(I,J))
          WRITE(6,320)
      320 FORMAT(1H0, ' CORE HAMILTONIAN ',/)
    C     CALL SCFOUT(0)
          RETURN
          END


          REAL FUNCTION FUNCJ*8(U,V,IND,JN,J1,J2,ASQ,ALPHA1,ALPHA2)
          IMPLICIT REAL*8(A-H,O-Z)
          IF(U.GE.1.D0) GO TO 10
          IF(U.LE.0.D0) GO TO 10
          IF(IND.EQ.2) GO TO 100
    C     FUNCTION FOR (I!M!J)
          F=U*(1.D0-U)*ASQ
          G=0.25D0*(ALPHA1**2/U+ALPHA2**2/(1.D0-U))
          X=2.D0*DSQRT(F*G)
          IF(JN.EQ.4) GO TO 20
          FUNCJ=0.25D0*(3.D0+3.D0*X+X*X)/(G**2.5*DEXP(X))
          GO TO 21
       20 CONTINUE
          FUNCJ=0.125D0*(15.D0+15.D0*X+6.D0*X*X+X**3)/(G**3.5*DEXP(X))
```

```
   21 CONTINUE
      FUNC=FUNCJ/(U**(FLOAT(J1)+0.5)*(1.D0-U)**(FLOAT(J2)+0.5))
      RETURN
  100 CONTINUE
C     FUNCTION FOR (I!M/R**3!J)
      IF(J1.GE.0) GO TO 11
      TAU=0.25D0*U*U*ASQ*(ALPHA1**2/U+ALPHA2**2/(1-U))
      SIGMA=(1-U)/U
      GO TO 12
   11 TAU=0.25D0*ASQ*(ALPHA1**2/U+ALPHA2**2/(1-U))
      SIGMA=0.D0
   12 X=2.D0*DSQRT((1.D0+SIGMA)*TAU)
      Y=2.D0*DSQRT(TAU*(SIGMA+V*V))
      IF(JN-2) 1,2,3
    1 FUNCT=(8.D0/3.D0)*V**4*DKBES(0,Y)
     1 +(2.D0/3.D0)*X*DKBES(1,X)/TAU
      GO TO 4
    2 FUNCT=(16.D0/15.D0)*V**6*DKBES(0,Y)
     1 +4.D0*X*DKBES(1,X)/(15.D0*TAU)+X*X*DKBES(2,X)/(3.D0*TAU*TAU)
      GO TO 4
    3 FUNCT=(32.D0/105.D0)*V**8*DKBES(0,Y)
     1 +8.D0*X*DKBES(1,X)/(105.D0*TAU) +2.D0*X*X*DKBES(2,X)/(15.D0*TAU*
     2 TAU) +X**3*DKBES(3,X)/(6.D0*TAU**3)
    4 CONTINUE
      IF(J1) 5,6,6
    5 IF(J2.GE.0) GO TO 7
      FUNC=FUNCT*U**(FLOAT(-J1)-0.5)*(1-U)**(FLOAT(-J2)-0.5)
      GO TO 8
    7 FUNC=FUNCT*U**(FLOAT(-J1)-0.5)/(1-U)**(FLOAT(J2)+0.5)
      GO TO 8
    6 IF(J2.GE.0) GO TO 9
      FUNC=FUNCT*(1-U)**(FLOAT(-J2)-0.5)/U**(FLOAT(J1)+0.5)
      GO TO 8
    9 FUNC=FUNCT/(U**(FLOAT(J1)+0.5)*(1-U)**(FLOAT(J2)+0.5))
    8 CONTINUE
      RETURN
   10 CONTINUE
      FUNC=0.D0
      RETURN
      END



      FUNCTION SS(NN1,LL1,MM,NN2,LL2,ALPHA,BETA)
      IMPLICIT REAL*8(A-H,O-Z)
C     PROCEDURE FOR CALCULATING REDUCED OVERLAP INTEGRALS
      COMMON/ARRAYS/S(80,80),Y(9,5,203),Z(17,45),XX(2900)
      COMMON/AUXINT/A(17),B(17)
      INTEGER ULIM
      N1=NN1
      L1=LL1
      M=MM
      N2=NN2
      L2=LL2
      P =(ALPHA + BETA)/2.D0
      PT=(ALPHA - BETA)/2.D0
      X = 0.D0
      M=IABS(M)
C     REVERSE QUANTUM NUMBERS IF NECESSARY
```

```
      IF((L2.LT.L1).OR.((L2.EQ.L1).AND.(N2.LT.N1))) GO TO 20
   10 GO TO 30
   20 K = N1
      N1 = N2
      N2= K
      K= L1
      L1= L2
      L2= K
      PT=-PT
   30 CONTINUE
      K = MOD((N1+N2-L1-L2),2)
C     FIND A AND B INTEGRALS
      NNN=N1+N2
      CALL AINTGS(P,NNN)
      CALL BINTGS(PT,NNN)
      IF((L1.GT.0).OR.(L2.GT.0)) GO TO 60
C     BEGIN SECTION USED FOR OVERLAP INTEGRALS INVOLVING S FUNCTIONS
C     FIND Z TABLE NUMBER L
   40 L = (90-17*N1+N1**2-2*N2)/2
      ULIM = N1+N2
      LLIM =0
      DO 50 I=LLIM,ULIM
      NNI1=N1+N2-I+1
   50 X=X+Z(I+1,L)*A(I+1)*B(NNI1)/2.D0
      SS=X
      GO TO 80
C     BEGIN SECTION USED FOR OVERLAPS INVOLVING NON-S FUNCTIONS
C     FIND Y TABLE NUMBER L
   60 L=(5-M)*(24-10*M+M**2)*(00 30*M+J*M**2)/120+
     1  (30-9*L1+L1**2-2*N1)*(28-9*L1+L1**2-2*N1)/8+
     2  (30-9*L2+L2**2-2*N2)/2
      LLIM=0
      DO 70 I=LLIM,8
      ULIM=4 - MOD(K+I,2)
      DO 70 J=LLIM,ULIM
      IIII=2*J+MOD(K+I,2)+1
   70 X=X+Y(I+1,J+1,L)*A(I+1)*B(IIII)
      L1MM=L1-M
      L2MM=L2-M
      L1PM=L1+M
      L2PM=L2+M
      SS=X*(FACT(M+1)/8.D0)**2*DSQRT(DFLOAT(2*L1+1)*FACT(L1MM)*
     1 DFLOAT(2*L2+1)*FACT(L2MM)/(4.D0*FACT(L1PM)*FACT(L2PM)))
   80 CONTINUE
      RETURN
      END



      SUBROUTINE HARMTR(T,MAXL,E)
      IMPLICIT REAL*8(A-H,O-Z)
      DIMENSION T(9,9),E(3)
      COST = E(3)
      IF((1.D0-COST**2).GT.0.0000000001) GO TO 20
   10 SINT = 0.D0
      GO TO 30
   20 SINT=DSQRT(1.D0-COST**2)
   30 CONTINUE
      IF(SINT.GT.0.000001D0)  GO TO 50
```

```
   40 COSP = 1.D0
      SINP = 0.D0
      GO TO 70
   50 COSP = E(1)/SINT
   60 SINP = E(2)/SINT
   70 CONTINUE
      DO 80 I=1,9
      DO 80 J=1,9
   80 T(I,J) = 0.D0
      T(1,1) = 1.D0
      IF(MAXL.GT.1) GO TO 100
   90 IF(MAXL.GT.0) GO TO 110
      GO TO 120
  100 COS2T = COST**2-SINT**2
      SIN2T = 2.D0*SINT*COST
      COS2P = COSP**2-SINP**2
      SIN2P = 2.D0*SINP*COSP
C     TRANSFORMATION MATRIX ELEMENTS FOR D FUNCTIONS
      SQRT3=DSQRT(3.D0)
      T(5,5) = (3.D0*COST**2-1.D0)/2.D0
      T(5,6) = -SQRT3*SIN2T/2.D0
      T(5,8) = SQRT3*SINT**2/2.D0
      T(6,5) = SQRT3*SIN2T*COSP/2.D0
      T(6,6) = COS2T*COSP
      T(6,7) = -COST*SINP
      T(6,8) = -T(6,5)/SQRT3
      T(6,9) = SINT*SINP
      T(7,5) = SQRT3*SIN2T*SINP/2.D0
      T(7,6) = COS2T*SINP
      T(7,7) = COST*COSP
      T(7,8) = -T(7,5)/SQRT3
      T(7,9) = -SINT*COSP
      T(8,5) = SQRT3*SINT**2*COS2P/2.D0
      T(8,6) = SIN2T*COS2P/2.D0
      T(8,7) = -SINT*SIN2P
      T(8,8) = (1.D0+COST**2)*COS2P/2.D0
      T(8,9) = -COST*SIN2P
      T(9,5) = SQRT3*SINT**2*SIN2P/2.D0
      T(9,6) = SIN2T*SIN2P/2.D0
      T(9,7) = SINT*COS2P
      T(9,8) = (1.D0+COST**2)*SIN2P/2.D0
      T(9,9) = COST*COS2P
  110 CONTINUE
      T(2,2) = COST*COSP
      T(2,3) = -SINP
      T(2,4) = SINT*COSP
      T(3,2) = COST*SINP
      T(3,3) = COSP
      T(3,4) = SINT*SINP
      T(4,2) = -SINT
      T(4,4) = COST
  120 CONTINUE
      RETURN
      END
```

```
      SUBROUTINE RELVEC(R,E,C1,C2)
      IMPLICIT REAL*8(A-H,O-Z)
      DIMENSION E(3),C1(3),C2(3)
      X = 0.D0
      DO 10 I=1,3
      E(I) = C2(I)-C1(I)
      X = X+E(I)**2
   10 CONTINUE
      R=DSQRT(X)
      DO 40 I=1,3
      IF (R.GT..000001D0) GO TO 30
   20 GO TO 40
   30 E(I) = E(I)/R
   40 CONTINUE
      RETURN
      END



      FUNCTION FACT(N)
      IMPLICIT REAL*8(A-H,O-Z)
      PRODT = 1.D0
   20 DO 30 I=1,N
   30 PRODT=PRODT*DFLOAT(I)
   40 FACT=PRODT
      RETURN
      END



      SUBROUTINE BINTGS(X,K)
      IMPLICIT REAL*8(A-H,O-Z)
      COMMON/AUXINT/A(17),B(17)
      ABSX=DABS(X)
      IF(ABSX.GT.3.D0) GO TO 120
   10 IF(ABSX.GT.2.D0) GO TO 20
   40 IF(ABSX.GT.1.D0) GO TO 50
   70 IF(ABSX.GT..5D0) GO TO 80
  100 IF(ABSX.GT..000001D0) GO TO 110
      GO TO 170
  110 LAST=6
      GO TO 140
   80 IF(K.LE.5) GO TO 120
   90 LAST=7
      GO TO 140
   50 IF(K.LE.7) GO TO 120
   60 LAST=12
      GO TO 140
   20 IF(K.LE.10) GO TO 120
   30 LAST=15
      GO TO 140
  120 EXPX=DEXP(X)
      EXPMX=1.D0/EXPX
      B(1)=(EXPX-EXPMX)/X
      DO 130 I=1,K
  130 B(I+1)=(DFLOAT(I)*B(I)+(-1.D0)**I*EXPX-EXPMX)/X
      GO TO 190
  140 Y=0.D0
      Y=2.D0
```

```
      DO 150 M=1,LAST
150   Y=Y+(-X)**M*(1.D0-(-1.D0)**(M+1))/(FACT(M)*DFLOAT(M+1))
      B(1)=Y
      DO 160 I=1,K
      Y=0.D0
      Y=Y+(1.D0-(-1.D0)**(I+1))/DFLOAT(I+1)
      DO 200 M=1,LAST
200   Y=Y+(-X)**M*(1.D0-(-1.D0)**(M+I+1))/(FACT(M)*DFLOAT(M+I+1))
160   B(I+1)=Y
      GO TO 190
170   B(1)=2.D0
      DO 180 I=1,K
180   B(I+1)=(1.D0-(-1.D0)**(I+1))/DFLOAT(I+1)
190   CONTINUE
      RETURN
      END



      SUBROUTINE AINTGS(X,K)
      IMPLICIT REAL*8(A-H,O-Z)
      COMMON/AUXINT/A(17),B(17)
      A(1)=DEXP(-X)/X
      DO 10 I=1,K
 10   A(I+1) =(A(I)*DFLOAT(I)+DEXP(-X))/X
      RETURN
      END



      REAL FUNCTION DGASPD*8(FUNC,N,M,C,D,E,F,IND,JN,J1,J2,ASQ,ALPHA1,
     1 ALPHA2)
      IMPLICIT REAL*8(A-H,O-Z)
      COMMON XXX(10)
      DIMENSION A(6,12),X(6,12),AA(24),U(24),BB(24),V(24)
     1,A1(19),A2(19),A3(19),A4(15)
     2,X1(19),X2(19),X3(19),X4(15)
      EQUIVALENCE (A(1),A1(1)),(A(20),A2(1)),(A(39),A3(1)),(A(58),A4(1))
     1,           (X(1),X1(1)),(X(20),X2(1)),(X(39),X3(1)),(X(58),X4(1))
      DATA   X1    /            0.86113631159405260D+00
     1,  0.96028985649753620D+00, 0.98156063424671920D+00
     2,  0.98940093499164990D+00, 0.99312859918509490D+00
     3,  0.99518721999702140D+00, 0.33998104358485630D+00
     4,  0.79666647741362670D+00, 0.90411725637047490D+00
     5,  0.94457502307323260D+00, 0.96397192727791380D+00,
     6,  0.97472855597130950D+00, 0.00000000000000000D-40
     7,  0.52553240991632900D+00, 0.76990267419430470D+00
     8,  0.86563120238783170D+00, 0.91223442825132590D+00
     9,  0.93827455200273280D+00, 0.00000000000000000D-40 /
      DATA   X2      /          0.18343464249564980D+00
     1,  0.58731795428661740D+00, 0.75540440835500300D+00
     2,  0.83911697182221880D+00, 0.86641552700440100D+00
     3,  0.00000000000000000D-40, 0.00000000000000000D-40
     4,  0.36783149899818020D+00, 0.61787624440264370D+00
     5,  0.74633190646015080D+00, 0.82000198597390290D+00
     6,  0.00000000000000000D-40, 0.00000000000000000D-40
     7,  0.12523340851146890D+00, 0.45801677765722740D+00
     8,  0.63605368072651500D+00, 0.74012419157855440D+00
     9,  0.00000000000000000D-40, 0.00000000000000000D-40 /
```

```
      DATA   X3       /              0.00000000000000000D-40
   1,   0.28160355077925890D+00,     0.51086700195082710D+00
   2,   0.64809365193697560D+00,     0.00000000000000000D-40
   3,   0.00000000000000000D-40,     0.00000000000000000D-40
   4,   0.95012509837637440D-01,     0.37370608871541960D+00
   5,   0.54542147138883950D+00,     0.00000000000000000D-40
   6,   0.00000000000000000D-40,     0.00000000000000000D-40
   7,   0.00000000000000000D-40,     0.22778585114164510D+00
   8,   0.43379350762604510D+00,     0.00000000000000000D-40
   9,   0.00000000000000000D-40,     0.00000000000000000D-40 /
      DATA   X4       /              0.51086700195082710D+00
   1,   0.76526521133497330D-01,     0.31504267969616340D+00
   2,   0.00000000000000000D-40,     0.00000000000000000D-40
   3,   0.00000000000000000D-40,     0.00000000000000000D-40
   4,   0.00000000000000000D-40,     0.19111886747361630D+00
   5,   0.00000000000000000D-40,     0.00000000000000000D-40
   6,   0.00000000000000000D-40,     0.00000000000000000D-40
   7,   0.00000000000000000D-40,     0.64056892862605620D-01 /
      DATA   A1       /              0.34785484513745390D+00
   1,   0.10122853629037630D+00,     0.47175336386511830D-01
   2,   0.27152459411754100D-01,     0.17614007139152120D-01
   3,   0.12341229799987200D-01,     0.65214515486254620D-01
   4,   0.22238103445337450D+00,     0.10693932599531840D+00
   5,   0.62253523938647890D-01,     0.40601429800386940D-01
   6,   0.28531388628933660D-01,     0.00000000000000000D-40
   7,   0.31370664587788730D+00,     0.16007832854334620D+00
   8,   0.95158511682492780D-01,     0.62672048334109070D-01
   9,   0.44277438817419810D-01,     0.00000000000000000D-40 /
      DATA   A2       /              0.36268378337836200D+00
   1,   0.20316742672306590D+00,     0.12462897125553390D+00
   2,   0.83276741576704750D-01,     0.59298584915436780D-01
   3,   0.00000000000000000D-40,     0.00000000000000000D-40
   4,   0.23349253653835480D+00,     0.14959598881657670D+00
   5,   0.10193011981724040D+00,     0.73346481411080310D-01
   6,   0.00000000000000000D-40,     0.00000000000000000D-40
   7,   0.24914704581340280D+00,     0.16915651939500250D+00
   8,   0.11819453196151840D+00,     0.86190161531953270D-01
   9,   0.00000000000000000D-40,     0.00000000000000000D-40 /
      DATA   A3       /              0.00000000000000000D-40
   1,   0.18260341504492360D+00,     0.13168863844917660D+00
   2,   0.97618652104113890D-01,     0.00000000000000000D-40
   3,   0.00000000000000000D-40,     0.00000000000000000D-40
   4,   0.18945061045506850D+00,     0.14209610931838210D+00
   5,   0.10744427011596560D+00,     0.00000000000000000D-40
   6,   0.00000000000000000D-40,     0.00000000000000000D-40
   7,   0.00000000000000000D-40,     0.14917298647260380D+00
   8,   0.11550566805372560D+00,     0.00000000000000000D-40
   9,   0.00000000000000000D-40,     0.00000000000000000D-40 /
      DATA   A4       /              0.13168863844917660D+00
   1,   0.15275338713072580D+00,     0.12167047292780340D+00
   2,   0.00000000000000000D-40,     0.00000000000000000D-40
   3,   0.00000000000000000D-40,     0.00000000000000000D-40
   4,   0.00000000000000000D-40,     0.12583745634682830D+00
   5,   0.00000000000000000D-40,     0.00000000000000000D-40
   6,   0.00000000000000000D-40,     0.00000000000000000D-40
   7,   0.00000000000000000D-40,     0.12793819534675220D+00 /
      I=N
      L=M
      IF(N-1) 600,100,100
```

5 PROGRAM LISTINGS

```
  100 IF(N-6) 200,200,600
  200 IF(M-1) 600,300,300
  300 IF(M-6) 400,400,600
  400 NN=4*N
      C1=(D+C)/2.D0
      C2=(D-C)/2.D0
      N2=N+N
      DO 1 J=1,N2
      NJ=NN-J+1
      U(J)=C1-C2*X(N,J)
      U(NJ)=C1+C2*X(N,J)
      AA(J)=A(N,J)
    1 AA(NJ)=A(N,J)
      IF(IND.EQ.1) GO TO 10
      MM=4*M
      D1=(F+E)/2.D0
      D2=(F-E)/2.D0
      M2=M+M
      DO 2 K=1,M2
      MK=MM-K+1
      V(K)=D1-D2*X(M,K)
      V(MK)=D1+D2*X(M,K)
      BB(K)=A(M,K)
    2 BB(MK)=A(M,K)
      S=0.0
      DO 3 J=1,NN
      DO 3 K=1,MM
    3 S=S+AA(J)*BB(K)*FUNC(U(J),V(K),IND,JN,J1,J2,ASQ,ALPHA1,ALPHA2)
      DGASPD=C2*D2*S
      N=I
      M=L
      RETURN
   10 CONTINUE
      S=0.D0
      DO 20 J=1,NN
   20 S=S+AA(J)*FUNC(U(J),U(J),IND,JN,J1,J2,ASQ,ALPHA1,ALPHA2)
      DGASPD=C2*S
      N=I
      RETURN
  600 WRITE(6,700) N,M
  700 FORMAT(1H0,'(SUBR. DGASPD) N=',I3,', M=',I3,',   N,M MUST BE LARGER
     1 THAN 0 AND SMALLER THAN 7.')
      N=6
      M=6
      GO TO 400
      END


      REAL FUNCTION DKBES * 8 (N,X)
      MODIFIED BESSEL FUNCTION OF THE SECOND KIND   -- DOUBLE --
      REAL * 8           X
      REAL * 8           XABS,RECX,HALFX,W,Y,SUM,EXPX,D75,D65,
     *                   AI0,AI1,AI2,AK0,AK1,AK2,NUMER,DENOM
     *            ,      Z
      REAL               XS,WS,YS,FK,C69,C173
      EQUIVALENCE (XABS,XS),(W,WS),(Y,YS)
      DATA D65,D75,C69,C173 / 1.0D65,1.0D75,69.0,173.0 /
      CALL OVERFL(KJ)
```

```
      XABS = X
      NABS=IABS(N)
      IF( XABS ) 1,1,3
    1 DKBES=D75
      WRITE(6,2) N,X
    2 FORMAT(1H ,5X,'THE ARGUMENT OF DKBES IS INVALID. N=',I10,'  ,   X=
     *',D23.16 )
      RETURN
C
    3 IF( NABS.GE.30000 )  GO TO 33
      IF(XS.LT.C173) GO TO 4
   33 DKBES=0.0D0
      RETURN
    4 RECX=1.0D0/XABS
      Z=RECX+RECX
C                          (Z=2/X)
      IF(XS .GT. 2.0)   GO TO 17
      HALFX=XABS*0.5D0
      W=DLOG(HALFX)
      IF(XS.GE.0.0001) GO TO 12
C                       X LESS THAN 0.0001
      Y=HALFX*HALFX
      AI0=1.0D0+Y
      AK0=-W*AI0-5.7721566490153286 D-1
     *          +4.22784335 D-1*Y
      IF(NABS) 6,5,6
    5 DKBES=AK0
      RETURN
    6 CONTINUE
      AK1=(RECX-HALFX*AK0)/AI0
      IF(NABS-1) 8,7,8
    7 AK2=AK1
      GO TO 26
    8 L=-C69/W
      IF(NABS .LE. L) GO TO 11
    9 DKBES=D75
      WRITE(6,10) N,X
   10 FORMAT(1H ,5X,'THE VALUE OF DKBES IS OVERFLOW. N=',I10,'  ,   X='
     *D23.16 )
      RETURN
   11 CONTINUE
      GO TO 22
C                       X  LESS THAN OR EQUAL TO 2.0,AND
C                       GREATER THAN OR EQUAL TO 0.0001.
C                       AT FIRST, I(0,X) AND I(1,X) ARE COMPUTED.
   12 CONTINUE
      AI2=0.0
      AI1=1.0D-75
      SUM=AI1
      K=18
      IF( XS .GE. 1.0 ) GO TO 13
      K=14
      IF( XS .GE. 0.1)  GO TO 13
      K=9
   13 FK=K
C                        RECURRENCE RELATION USED
C                        IN DESCENDING ORDER.
      AI0=( AI1*Z )*FK+AI2
      K=K-1
```

```
         IF(K) 14,15,14
      14 SUM=SUM+AIO
         AI2=AI1
         AI1=AIO
         GO TO 13
      15 SUM=SUM+SUM+AIO
         EXPX=DEXP( XABS )
         AIO=( AIO/SUM )*EXPX
C                         THE VALUE OF I(O,X) HAS BEEN OBTAINED.
C                         COMPUTE K(O,X) USING POWER SERIES EXPANSION
         Y=HALFX*HALFX
C
         AKO = ( ( ( 1.1  E-17 * YS + 1.533  E-15 ) * YS
        *             +1.78593E-13)* YS + 1.709994E-11)*YS
         AKO = ( ( AKO + 1.31674867      D-9 ) * Y + 7.9350965213  D-8 ) * Y
         AKO = -W * AIO + ( ( ( ( ( ( AKO
        A + 3.61262410320       D-6 ) * Y + 1.18480393641097  D-4 )* Y
        B + 2.61478761880521    D-3 ) * Y + 3.489215745643890 D-2 )* Y
        C + 2.30696083774611679D-1 ) * Y + 4.22784335098467140D-1 )* Y
        D - 5.77215664901532860D-1
C
         IF( NABS ) 16,5,16
      16 CONTINUE
C                         COMPUTE K(1,X) USING LOMMEL'S RELATION
         AI1=(AI1/SUM)* EXPX
         AK1=(RECX-AKO*AI1)/AIO
         IF( NABS .EQ. 1 ) GO TO 7
         GO TO 22
C                         X GREATER THAN 2.0
C                         K(O,X) AND K(1,X) ARE COMPUTED BY MEANS
C                         OF RATIONAL EXPRESSIONS
      17 Y=RECX*0.5D0
C                         ( Y=1/2X )
         W= DEXP( - XABS) * DSQRT( 3.14159265358979D0*Y )
         IF( NABS .EQ. 1 )  GO TO 29
         IF( XS .GE. 6.0 ) GO TO 18
         NUMER = ( ( ( ( ( ( (
        A    9.607359468936920D-1   * Y + 2.529252282967791D+2 ) * Y
        B + 7.970033517428499D+3 ) * Y + 7.569208233441645D+4 ) * Y
        C + 3.057001687861112D+5 ) * Y + 6.316930369273204D+5 ) * Y
        D + 7.490317987301367D+5 ) * Y + 5.503149540522960D+5 ) * Y
         NUMER = ( (    NUMER
        F + 2.642871289210102D+5 ) * Y + 8.617173613052524D+4 ) * Y
C
         NUMER = ( ( ( ( ( ( (             NUMER
        G + 1.959077941045564D+4 ) * Y + 3.161402444515176D+3 ) * Y
        H + 3.658576787591567D+2 ) * Y + 3.045630347257958D+1 ) * Y
        I + 1.815056853732798D 0 ) * Y + 7.630840968696385D-2 ) * Y
        J + 2.198080790671052D-3 ) * Y
         NUMER = ( ( (    NUMER
        K + 4.108400060979278D-5 ) * Y + 4.468603071316900D-7 ) * Y
        L + 2.138087593931531D-9 ) * Y
C
         DENOM = ( ( ( ( ( ( (                 Y
        A + 3.555555555555556D+2 ) * Y + 1.513244444444444D+4 ) * Y
        B + 1.956717714285714D+5 ) * Y + 1.079473194053918D+6 ) * Y
        C + 3.045125484052925D+6 ) * Y + 4.914134724130594D+6 ) * Y
        D + 4.892294125356680D+6 ) * Y
         DENOM = ( ( (    DENOM
```

```
            E + 3.168987751788133D+6  ) * Y + 1.388013537038700D+6  ) * Y
            F + 4.227486032369927D+5  ) * Y
C
             DENOM = ( ( ( ( ( ( (                    DENOM
            G + 9.133105119891823D+4  ) * Y + 1.418093261050334D+4  ) * Y
            H + 1.593555554804435D+3  ) * Y + 1.296908735314483D+2  ) * Y
            I + 7.594657346992128D 0  ) * Y + 3.149536504573292D-1  ) * Y
            J + 8.975302543644857D-3  ) * Y
             DENOM = ( (       DENOM
            K + 1.663376533185147D-4  ) * Y + 1.797062622699450D-6  ) * Y
            L + 8.552350375726116D-9
C
          GO TO 19
C
       18 CONTINUE
            NUMER = ( ( ( ( ( ( (
            A    9.344272845910382D-1      * Y + 7.602616403443910D+1  ) * Y
            B + 7.395743295243645D+2  ) * Y + 2.143299351353469D+3  ) * Y
            C + 2.591022778768068D+3  ) * Y + 1.352060383786633D+3  ) * Y
            D + 5.070112211483554D+2  ) * Y + 9.496661840062622D+1  ) * Y
             NUMER = ( ( ( (    NUMER
            E + 1.035232548375690D+1  ) * Y + 6.417643499456601D-1  ) * Y
            F + 2.076251506438793D-2  ) * Y + 2.696430527842588D-4  ) * Y
C
             DENOM = ( ( ( ( ( ( (              Y + 128.0D0) * Y
            A + 1.952426666666667D+3  ) * Y + 8.925379047619048D+3  ) * Y
            B + 1.700072199546485D+4  ) * Y + 1.598598652282462D+4  ) * Y
            C + 8.186476153700395D+3  ) * Y + 2.418159110016117D+3  ) * Y
             DENOM =( ( ( (    DENOM
            D + 4.239448497375430D+2  ) * Y + 4.421129278670809D+1  ) * Y
            E + 2.659325881907253D 0  ) * Y + 8.426345399508084D-2  ) * Y
            F + 1.078572211137035D-3
C
       19 AKO = W * ( 1.0D0 - NUMER / DENOM )
           IF( NABS .EQ. 0 )  GO TO 5
       29 CONTINUE
C                                  X IS GREATER    THAN 2.0
C                                  N IS GREATER    THAN OR EQUAL TO 1
C                                  RATIONAL APPROXIMATION IS USED TO COMPUTE
C                                  K(1,X)
           IF( XS .GE. 6.0 )  GO TO 20
C
             NUMER = ( ( ( ( ( ( (
            A    4.612351477146149D+1      * Y + 5.662820251461119D+3  ) * Y
            B + 1.267729206953107D+5  ) * Y + 9.840279611657081D+5  ) * Y
            C + 3.490851284631075D+6  ) * Y + 6.612450857584894D+6  ) * Y
            D + 7.385469550146852D+6  ) * Y + 5.204090498624053D+6  ) * Y
             NUMER = ( (    NUMER
            E + 2.426670289964806D+6  ) * Y + 7.748520914563417D+5  ) * Y
C
             NUMER = ( ( ( ( ( ( (               NUMER
            F + 1.735592435557683D+5  ) * Y + 2.771333212784498D+4  ) * Y
            G + 3.183321561628970D+3  ) * Y + 2.636216725447433D+2  ) * Y
            H + 1.565459511840883D+1  ) * Y + 6.565891381439845D-1  ) * Y
            I + 1.888507334035547D-2  ) * Y
             NUMER = ( ( ( NUMER
            J + 3.526831364930226D-4  ) * Y + 3.834684961199849D-6  ) * Y
            K + 1.834777493397057D-8  ) * Y
```

5 PROGRAM LISTINGS

```
C
      DENOM = ( ( ( ( ( ( (                Y
    A + 640.0D0              )      + 3.242666666666667D+4 ) * Y
    B + 4.565674666666667D+5 ) * Y + 2.649616021768707D+6 ) * Y
    C + 7.729933921057426D+6 ) * Y + 1.277675028273955D+7 ) * Y
    D + 1.295019033182651D+7 ) * Y
      DENOM = ( ( (    DENOM           + 8.506230281115516D+6 ) * Y
    E + 3.767465314819330D+6 ) * Y + 1.157963565388285D+6 ) * Y
C
      DENOM = ( ( ( ( ( ( (           DENOM
    F + 2.520737013090144D+5 ) * Y + 3.939147947362039D+4 ) * Y
    G + 4.450965515143424D+3 ) * Y + 3.639711612011614D+2 ) * Y
    H + 2.140312525061418D+1 ) * Y + 8.908688970078743D-1 ) * Y
    I + 2.547045316439757D-2 ) * Y
      DENOM = ( ( (   DENOM
    J + 4.734225517526958D-4 ) * Y + 5.128203094044773D-6 ) * Y
    K + 2.446369991196075D-8
C
      GO TO 21
C
   20 CONTINUE
      NUMER = ( ( ( ( ( ( (
    A   2.727430674283096D+1       * Y + 1.138548998351567D+3 ) * Y
    B + 8.476942664996772D+3 ) * Y + 2.136946909366115D+4 ) * Y
    C + 2.388756033308294D+4 ) * Y + 1.367517138363589D+4 ) * Y
    D + 4.350909161812486D+3 ) * Y + 8.026823492415450D+2 ) * Y
      NUMER = ( ( ( (  NUMER
    E + 8.677093809042819D+1 ) * Y + 5.356622184821665D 0 ) * Y
    F + 1.730209588698993D-1 ) * Y + 2.247025439868823D-3 ) * Y
      DENOM = ( ( ( ( ( (         Y  +  2.304 D+2  ) * Y
    A + 4.183771428571428D+3 ) * Y + 2.082588444444444D+4 ) * Y
    B + 4.172904489795918D+4 ) * Y + 4.057981194255480D+4 ) * Y
    C + 2.128483799962103D+4 ) * Y + 6.401009408866192D+3 ) * Y
      DENOM = ( ( ( (   DENOM
    D + 1.137957228242879D+3 ) * Y + 1.200020804210648D+2 ) * Y
    E + 7.284240459137258D 0 ) * Y + 2.325671330264231D-1 ) * Y
    F + 2.996033919825096D-3
C
   21 CONTINUE
      AK1 = W * ( 1.0D0 + NUMER/DENOM )
      IF( NABS .EQ. 1 )  GO TO 7
C                           RECURRENCE RELATION IS USED IN ASCENDING
C                           ORDER 10 CALCULATE K(N,X)
   22 CONTINUE
      K=1
   23 FK=K
      IF( DABS( AK1 ).GE. D65 ) GO TO 25
      AK2=AK1*Z*FK+AK0
   24 K=K+1
      IF( K .GE. NABS ) GO TO 26
      AK0=AK1
      AK1=AK2
      GO TO 23
   25 W=AK1*1.0D-10
      Y=AK0*1.0D-10
      AK2=W*Z*FK+Y
      IF(AK2.GE.D65) GO TO 9
      AK2=AK2/1.0D-10
      GO TO 24
```

```
C                         VALUE OF K(NABS,X) HAS BEEN OBTAINED
C                         IN AK2
   26 CONTINUE
      IF(X)  27, 28, 28
   27 K=NABS/2
      IF( K+K .NE. NABS )  AK2=-AK2
   28 DKBES=AK2
      RETURN
      END


      SUBROUTINE COEFFT
      IMPLICIT REAL*8 (A-H,O-Z)
      COMMON/ARRAYS/S(80,80),Y(9135),Z(765),XX(2900)
      DO 1 I=1,9135
    1 Y(I)=0.0D0
      DO 2 I=1,765
    2 Z(I)=0.0D0
C     LOAD NON-ZERO Y COEFFICIENTS
      Y(7039)= 64.     DO              Y(2720)= -96.    DO
      Y(7040)= 64.     DO              Y(2729)= 48.     DO
      Y(7049)= -64.    DO              Y(2703)= -48.    DO
      Y(7032)= -128.   DO              Y(2712)= -48.    DO
      Y(7041)= -64.    DO              Y(2721)= 96.     DO
      Y(7033)= -128.   DO              Y(2704)= -48.    DO
      Y(7042)= 128.    DO              Y(2713)= 48.     DO
      Y(7025)= 64.     DO              Y(2722)= 48.     DO
      Y(7034)= 120.    DO              Y(2731)= -48.    DO
      Y(7026)= 64.     DO              Y(2705)= 96.     DO
      Y(7035)= -64.    DO              Y(2714)= -48.    DO
      Y(7027)= -64.    DO              Y(2723)= -48.    DO
      Y(6904)= -96.    DO              Y(2706)= 48.     DO
      Y(6913)= 32.     DO              Y(2715)= -96.    DO
      Y(6896)= -192.   DO              Y(2724)= 48.     DO
      Y(6905)= 192.    DO              Y(2707)=-48.     DO
      Y(6906)= 288.    DO              Y(2716)= 48.     DO
      Y(6915)= -96.    DO              Y(5329)= 64.     DO
      Y(6889)= 192.    DO              Y(5322)= -128.   DO
      Y(6907)= -192.   DO              Y(5340)= -64.    DO
      Y(6890)= 96.     DO              Y(5315)= 64.     DO
      Y(6899)= -288.   DO              Y(5333)= 128.    DO
      Y(6891)= -192.   DO              Y(5326)= -64.    DO
      Y(6900)= 192.    DO              Y(5185)= -96.    DO
      Y(6892)= -32.    DO              Y(5194)= -32.    DO
      Y(6901)= 96.     DO              Y(5186)= -96.    DO
      Y(2854)= -16.    DO              Y(5195)= 64.     DO
      Y(2863)= 16.     DO              Y(5204)= 32.     DO
      Y(2847)= 32.     DO              Y(5178)= 96.     DO
      Y(2856)= -16.    DO              Y(5187)= 32.     DO
      Y(2865)= -16.    DO              Y(5196)= 64.     DO
      Y(2840)= -16.    DO              Y(5179)= 96.     DO
      Y(2849)= -16.    DO              Y(5188)= -32.    DO
      Y(2858)= 32.     DO              Y(5197)= 32.     DO
      Y(2842)= 16.     DO              Y(5206)= -96.    DO
      Y(2851)= -16.    DO              Y(5180)= -64.    DO
      Y(2710)= 48.     DO              Y(5189)= -32.    DO
      Y(2719)= -48.    DO              Y(5198)= -96.    DO
      Y(2711)= 48.     DO              Y(5181)= -32.    DO
```

```
Y(5190)= -64.    DO
Y(5199)= 96.     DO
Y(5182)= -32.    DO
Y(5191)= 96.     DO
Y(4375)= -144.   DO
Y(4384)= 96.     DO
Y(4393)= -16.    DO
Y(4368)= 144.    DO
Y(4386)= -48.    DO
Y(4395)= 96.     DO
Y(4370)= -96.    DO
Y(4379)= 48.     DO
Y(4397)= -144.   DO
Y(4372)= 16.     DO
Y(4381)= -96.    DO
Y(4390)= 144.    DO
Y(1900)= 144.    DO
Y(1909)= -144.   DO
Y(1893)= -144.   DO
Y(1920)= 144.    DO
Y(1895)= 144.    DO
Y(1922)= -144.   DO
Y(1906)= -144.   DO
Y(1915)= 144.    DO
Y( 955)= -16.    DO
Y( 964)= 32.     DO
Y( 973)= -16.    DO
Y( 948)= 16.     DO
Y( 966)= -48.    DO
Y( 975)= 32.     DO
Y( 950)= -32.    DO
Y( 959)= 48.     DO
Y( 977)= -16.    DO
Y( 952)= 16.     DO
Y( 961)= -32.    DO
Y( 970)= 16.     DO
Y(8155)= 64.     DO
Y(8156)= -64.    DO
Y(8165)= -64.    DO
Y(8148)= -64.    DO
Y(8157)= 64.     DO
Y(8149)= 64.     DO
Y(8158)= 64.     DO
Y(8150)= -64.    DO
Y(8020)= -96.    DO
Y(8029)= 32.     DO
Y(8021)= 128.    DO
Y(8013)= 96.     DO
Y(8031)= -96.    DO
Y(8014)= -128.   DO
Y(8015)= -32.    DO
Y(8024)= 96.     DO
Y(7084)= -64.    DO
Y(7076)= -128.   DO
Y(7085)= 64.     DO
Y(7086)= 128.    DO
Y(7069)= 128.    DO
Y(7070)= 64.     DO
Y(7079)= -128.   DO

Y(7071)= -64.    DO
Y(3205)= -16.    DO
Y(3214)= 16.     DO
Y(3206)= 16.     DO
Y(3215)= -16.    DO
Y(3198)= 16.     DO
Y(3216)= -16.    DO
Y(3199)= -16.    DO
Y(3217)= 16.     DO
Y(3200)= -16.    DO
Y(3209)= 16.     DO
Y(3201)= 16.     DO
Y(3210)= -16.    DO
Y(7579)= 64.     DO
Y(7580)= -64.    DO
Y(7572)= -128.   DO
Y(7573)= 128.    DO
Y(7565)= 64.     DO
Y(7566)= -64.    DO
Y(5680)= 64.     DO
Y(5681)= -64.    DO
Y(5673)= -64.    DO
Y(5691)= -64.    DO
Y(5674)= 64.     DO
Y(5692)= 64.     DO
Y(5684)= 64.     DO
Y(5685)= -64.    DO
Y(7435)= -96.    DO
Y(7444)= 32.     DO
Y(7436)= -96.    DO
Y(7445)= 160.    DO
Y(7428)= 96.     DO
Y(7437)= 128.    DO
Y(7446)= -96.    DO
Y(7429)= 96.     DO
Y(7438)= -128.   DO
Y(7447)= -96.    DO
Y(7430)= -160.   DO
Y(7439)= 96.     DO
Y(7431)= -32.    DO
Y(7440)= 96.     DO
Y(5545)= -96.    DO
Y(5554)= 32.     DO
Y(5546)= 32.     DO
Y(5555)= 32.     DO
Y(5538)= 96.     DO
Y(5556)= 32.     DO
Y(5539)= -32.    DO
Y(5557)= -96.    DO
Y(5540)= -32.    DO
Y(5549)= -32.    DO
Y(5541)=-32.     DO
Y(5550)= 96.     DO
Y(3070)= 48.     DO
Y(3079)= -48.    DO
Y(3071)= -48.    DO
Y(3080)= 48.     DO
Y(3063)= -48.    DO
Y(3081)= 48.     DO
```

```
      Y(3082)= -48.    DO
      Y(3064)= 48.     DO
      Y(3065)= 48.     DO
      Y(3074)= -48.    DO
      Y(3066)= -48.    DO
      Y(3075)= 48.     DO
      Y(8200)= -64.    DO
      Y(8201)= 64.     DO
      Y(8193)= 64.     DO
      Y(8194)= -64.    DO
      Y(7615)= -64.    DO
      Y(7616)= -64.    DO
      Y(7625)= 64.     DO
      Y(7608)= 64.     DO
      Y(7617)= 64.     DO
      Y(7609)= 64.     DO
      Y(7618)= -64.    DO
      Y(7610)= -64.    DO
      Y(3250)= 16.     DO
      Y(3259)= -16.    DO
      Y(3243)= -16.    DO
      Y(3261)= 16.     DO
      Y(3245)= 16.     DO
      Y(3254)= -16.    DO
      Y(5725)= -64.    DO
      Y(5718)= 64.     DO
      Y(5736)= 64.     DO
      Y(5729)= -64.    DO
C     LOAD NON-ZERO Z COEFFICIENTS
      Z(341 )= -1.     DO
      Z(343 )= 3.      DO
      Z(345 )= -3.     DO
      Z(347 )= 1.      DO
      Z(664 )= -1.     DO
      Z(665 )= 5.      DO
      Z(666 )= -10.    DO
      Z(667 )= 10.     DO
      Z(668 )= -5.     DO
      Z(669 )= 1.      DO
      Z(154 )= -1.     DO
      Z(156 )= 5.      DO
      Z(158 )= -10.    DO
      Z(160 )= 10.     DO
      Z(162 )= -5.     DO
      Z(164 )= 1.      DO
      Z(222 )= -1.     DO
      Z(223 )= 1.      DO
      Z(224 )= 4.      DO
      Z(225 )= -4.     DO
      Z(226 )= -6.     DO
      Z(227 )= 6.      DO
      Z(228 )= 4.      DO
      Z(229 )= -4.     DO
      Z(230 )= -1.     DO
      Z(231 )= 1.      DO
      Z(307 )= -1.     DO
      Z(308 )= 2.      DO
      Z(309 )= 2.      DO
      Z(310 )= -6.     DO
      Z(312 )= 6.      DO
      Z(313 )= -2.     DO
      Z(314 )= -2.     DO
      Z(315 )= 1.      DO
      Z(409 )= -1.     DO
      Z(410 )= 3.      DO
      Z(411 )= -1.     DO
      Z(412 )= -5.     DO
      Z(413 )= 5.      DO
      Z(414 )= 1.      DO
      Z(415 )= -3.     DO
      Z(416 )= 1.      DO
      Z(528 )= -1.     DO
      Z(529 )= 4.      DO
      Z(530 )= -5.     DO
      Z(532 )= 5.      DO
      Z(533 )= -4.     DO
      Z(534 )= 1.      DO
      Z(562 )= -1.     DO
      Z(563 )= 2.      DO
      Z(565 )= -2.     DO
      Z(566 )= 1.      DO
      Z(732 )= -1.     DO
      Z(733 )= 1.      DO
      Z(545 )= 1.      DO
      Z(546 )= -3.     DO
      Z(547 )= 2.      DO
      Z(548 )= 2.      DO
      Z(549 )= -3.     DO
      Z(550 )= 1.      DO
      Z(579 )= 1.      DO
      Z(580 )= -1.     DO
      Z(581 )= -1.     DO
      Z(582 )= 1.      DO
      Z(596 )= -1.     DO
      Z(598 )= 1.      DO
      Z(443 )= -1.     DO
      Z(444 )= 1.      DO
      Z(445 )= 2.      DO
      Z(446 )= -2.     DO
      Z(447 )= -1.     DO
      Z(448 )= 1.      DO
      Z(698 )= -1.     DO
      Z(699 )=3.       DO
      Z(700 )= -3.     DO
      Z(701 )= 1.      DO
      Z(324 )= 1.      DO
      Z(325 )= -1.     DO
      Z(326 )= -3.     DO
      Z(327 )= 3.      DO
      Z(328 )= 3.      DO
      Z(329 )= -3.     DO
      Z(330 )= -1.     DO
      Z(331 )= 1.      DO
      Z(460 )= 1.      DO
      Z(462 )= -2.     DO
      Z(464 )= 1.      DO
      RETURN
      END
```

5.B A PROGRAM FOR CALCULATING NUCLEAR SHIELDING BY MEANS OF THE INDO/S-SOS PROCEDURE

The program is written for a CDC 7600 system in FORTRAN IV. As it stands the program is capable of calculating the shielding, without configuration interaction (CI), for carbon, nitrogen, oxygen and fluorine nuclei. The SOS perturbation procedure[4,5] is added to the basic CNDO/INDO program, QCPE 174, and the INDO/S parameterization scheme is employed.[6-8] The program also gives the AEE results for the paramagnetic shielding term.

The maximum number of atoms, NA, allowed is 20 and the maximum number of orbitals, N, is 60.

5.B.1 Description of the Input Data

First three cards	Comment cards for alphanumeric information to be printed at the head of the output.
Fourth card	Number of orbitals employed, N, FORMAT (I4). Number of occupied orbitals, NOCC, FORMAT (I4). Number of atoms considered, NA, FORMAT (I4). To use subroutine CONVERG, put ISTOP $\neq 0$, otherwise ISTOP $= 0$, FORMAT (I4). To obtain molecular orbitals and energies, put ISUB > 0, otherwise ISUB ≤ 0, FORMAT (I4). If both the CI matrix and the CI states are required, put IDUMB > 100. If only the CI states are required put $100 <$ IDUMB > 0, otherwise IDUMB $= 0$, FORMAT (I4).
NA cards	One for each atom containing its x, y and z coordinates, FORMAT (F10.6) and its atomic number, FORMAT (I4).
One card	Number of the atom(s) for which a study of shielding contributions from various electronic transitions is required FORMAT (20I2)
One card	Type of molecular orbital, whether (π, σ or n) PI, SIGMA or NON-BONDING FORMAT (60A1). This and the previous card are left blank if contributions of electronic transitions to shielding are not required.
Final card	END, FORMAT (A3), terminated data deck.

5.B.2 The Program

```
C     NUCLEAR MAGNETIC SHIELDING CALCULATIONS WITHOUT CI
C     THE FOLLOWING INPUT CARDS ARE REQUIRED,
C     1. THREE COMMENT CARDS ANY INFORMATION ALL 80 COLUMNS.
C
C     2. A. THE NUMBER OF ORBITALS FORMAT I4,N.
C        B. THE NUMBER OF OCCUPIED ORBITALS FORMAT I4,NOCC.
C        C. THE NUMBER OF CENTERS FORMAT I4,NA.
C        D. ISTOP#0 IF SUBROUTINE CONVERG IS TO BE USED
C           OTHERWISE ISTOP = 0 FORMAT I4
C        E. ISUB IS GREATER THAN ZERO IF MOLECULAR ORBITALS AND
C           ENERGIES ARE REQUIRED,
C           OTHERWISE ISUB IS LESS OR EQUAL TO ZERO FORMAT I4
C        F. IDUMB >100 IF BOTH CI MATRIX @ CI STATES ARE REQUIRED .
C           100< IDUMB >0 IF ONLY CI STATES REQUIRED. ELSE IDUMB=0 .
C     3. ONE CARD FOR EACH ATOM WITH ITS X,Y, AND Z COORDINATES
C        FORMAT F10.6 , AND ITS ATOMIC NUMBER IN FORMAT I4 .
C        THERE ARE NA OF THESE CARDS.
C
C     4. ONE CARDS FOR IATOMS(20)   20I2
C        NUMBER OF THE ATOM FOR WHICH ASTUDY OF CONTRIBUTIONS
C        FROM VARIOUS ELECT. TRANSITIONS IS REQUIRED
C     5. ITYPE(60) , 60A1  TYPE OF MO'S WETHERE PI(P) , SIGMA(S)
C        OR NONBONDING(N)    THESE TWO CARDS ARE REPLACED BY BLANCKS
C        IF THIS STUDY IS NOT REQUIRED ************
C     6. DATA DECK IS TERMINATED BY A CARD CONTAINING END IN THE
C        FIRST THREE COLUMNS    I.E.    A3
C
      COMMON /ONE/ N,NA,NOCC,ICHGE,IX,IY,I,J,KI,KJ,ISUB,IDUMB,ISTOP,
     1E,TCONV,MU,MO
      COMMON/TWO/ AII(100),AIII(100),XC(35),YC(35),ZC(35),A(100,100),
     1           BETAO(100,100),C(100,100)
      COMMON/THREE/ NAT(35),S(15),ZN(35),ZCHG(35),Q(35)
      COMMON/FOUR/ P(35,35),PI(35,35)
      COMMON/FIVE/ PO(100,100),POI(100,100)
      COMMON/SIX/ U(35,3),R(35,35),BETA(35,35),GAMMA(35,35)
      COMMON/GEOD/ EA(3),T(9,9),TS(9,9),SA(20)
      IX = 5
      IY = 6
  900 WRITE(IY,100)
      CALL TIME (TM)
      WRITE (IY,102) TM
  100 FORMAT(1H1)
  101 FORMAT(I4)
  102 FORMAT (/13H THE TIME IS ,A8)
  103 FORMAT (/25H SUBROUTINE CONVERGE USED,/)
      CALL INPUT
      IF (ISTOP.GT.0) WRITE (IY,103)
      CALL REPLSN
      CALL BET1
      CALL INDOMO
      CALL BET2
      CALL REPS
      CALL CIMAT
      CALL SHIFT1
      CALL TIME(TM)
      WRITE (IY,102) TM
      CONTINUE
```

5 PROGRAM LISTINGS

```
      STOP
      END

      SUBROUTINE INPUT
      COMMON /ONE/ N,NA,NOCC,ICHGE,IX,IY,I,J,KI,KJ,ISUB,IDUMB,ISTOP,
     1E,TCONV,MU,MO
      COMMON/TWO/ AII(100),AIII(100),XC(35),YC(35),ZC(35),A(100,100),
     1            BETAO(100,100),C(100,100)
      COMMON/THREE/ NAT(35),S(15),ZN(35),ZCHG(35),Q(35)
      COMMON/FOUR/ P(35,35),PI(35,35)
      COMMON/FIVE/ PO(100,100),POI(100,100)
      COMMON/SIX/ U(35,3),R(35,35),BETA(35,35),GAMMA(35,35)
      COMMON/SEVEN/ KHARGE,IMULT,NFIRST(35),NLAST(35)
      DIMENSION IDENT(20)
      DATA IEND/4HEND /
  101 FORMAT(20A4)
  102 FORMAT (6I4)
  103 FORMAT(I4,3(3X,F12.7))
  104 FORMAT(5X,60H ATOM     X COORDINATE    Y COORDINATE    Z COORDINATE
     1AT. NO., 5X, 14HCORE INTEGRALS/)
  105 FORMAT(6X,I3,3X,F10.6,5X,F10.6,5X,F10.6,6X,I4,6X,F7.3,3X,F7.3,
     1       3X,F7.3)
  106 FORMAT(26X,6HCHARGE,2X,5HIMULT,2X,4H  NA,2X,5HISTOP,
     1 12X,4HISUB,2X,5HIDUMB/)
  108 FORMAT(26X,20A4)
  109 FORMAT (25X,I4,2X,I4,2X,I4,2X,I4,4X,I4,3X,I4 )
      DO 201 I = 1,3
      READ(IX,101) IDENT
      J = 3
      IF(IDENT(1).EQ.IEND)GO TO 99
  201 WRITE(IY,108) IDENT
C     N IS THE TOTAL NUMBER OF VALENCE SHELL ORBITALS
C     NOCC IS  NUMBER OF ELECTRONS/2
C     NA IS THE NUMBER OF ATOMS
C     E AND TCONV ARE CONVERGENCE CRITERIA
      TCONV=0.00001
      E = 0.0001
      READ(IX,102) KHARGE,IMULT,NA,ISTOP,ISUB,IDUMB
      WRITE (IY,106)
      WRITE(IY,109) KHARGE,IMULT,NA,ISTOP,ISUB,IDUMB
      DO 202 I = 1,NA
  202 READ(IX,103) NAT(I),XC(I),YC(I),ZC(I)
      WRITE(IY,104)
      DO 203 I = 1,NA
C     CONSTRUCTIONS OF INTERATOMIC DISTANCE MATRIX
      DO 203 J = 1,NA
      XQ = (XC(I) - XC(J))**2
      YQ = (YC(I) - YC(J))**2
      ZQ =(ZC(I) - ZC(J))**2
  203 R(I,J) = SQRT(XQ + YQ + ZQ)
      DO 199 I = 1,N
  199 AII(I) = 1.0
C     BETO ARE THE ATOMIC RESONANCE INTEGRALS
C     UUSO ARE THE CORE INTEGRALS FOR S ORBITALS
C     UUPO ARE THE CORE INTEGRALS FOR P ORBITALS
C     UUDO ARE THE CORE INTEGRALS FOR D ORBITALS
C     GAMO ARE THE ONE-CENTER ELECTRON REPULSION INTEGRALS
C     ZCHG IS THE CORE CHARGE
C     ZN IS THE SLATER EXPONENT
```

```
      DO 211 I = 1,NA
      IF(NAT(I) - 1) 207, 206, 207
  206 ZCHG(I) = 1.0
      ZN(I) = 1.2
      BETA(I,I) = -12.0/27.204
      U(I,1)=14.342
      U(I,2) = 0.0
      U(I,3) = 0.0
      GAMMA(I,I)=12.484
      GO TO 211
  207 IF(NAT(I).GT.10) GO TO 230
      ZX = NAT(I)
      ZCHG(I) = ZX - 2.0
      NNN = NAT(I)-4
      GO TO(300,208,209,210,212),NNN
  300 ZN(I) = 1.30
      BETA(I,I) = -7.0/27.204
      U(I,1) = 19.188
      U(I,2) = 8.022
      U(I,3) = 0.0
      GAMMA(I,I) = 8.30
      GO TO 211
  208 ZN(I) = 1.625
      BETA(I,I) = -17.5/27.204
      U(I,1)=29.924
      U(I,2)=11.612
      U(I,3) = 0.0
      GAMMA(I,I)=10.333
      GO TO 211
  209 ZN(I) = 1.95
      BETA(I,I) = -26.0/27.204
      U(I,1)=40.718
      U(I,2)=16.228
      U(I,3) = 0.0
      GAMMA(I,I)=11.308
      GO TO 211
  210 ZN(I) = 2.275
      BETA(I,I)=-31.0/27.204
      U(I,1)=50.78
      U(I,2)=18.222
      U(I,3) = 0.0
      GAMMA(I,I)=15.13
      GO TO 211
  212 ZN(I) = 2.60
      BETA(I,I)=-35.0/27.204
      U(I,1) = 64.544
      U(I,2) = 22.160
      U(I,3) = 0.0
      GAMMA(I,I)=18.0
      GO TO 211
  230 CONTINUE
      ZX = NAT(I)
      ZCHG(I) = ZX - 10.0
      MMM = NAT(I) - 12
      GO TO (400,401,402,403,404),MMM
  400 ZN(I) = 1.167
      BETA(I,I) = -4.0/27.204
      U(I,1) = 15.542
      U(I,2) = 5.990
```

5 PROGRAM LISTINGS

```
      U(I,3) = 0.448
      GAMMA(I,I) = 4.0
      GO TO 211
  401 ZN(I) = 1.38
      BETA(I,I) = -10.0/27.204
      U(I,1) = 20.066
      U(I,2) = 8.266
      U(I,3) = 0.674
      GAMMA(I,I) = 7.015
      GO TO 211
  402 ZN(I) = 1.6
      BETA(I,I) = -12.0/27.204
      U(I,1) = 28.066
      U(I,2) = 10.928
      U(I,3) = 1.0
      GAMMA(I,I) = 9.886
      GO TO 211
  403 ZN(I) = 1.82
      BETA(I,I) = -15.0/27.204
      U(I,1) = 35.300
      U(I,2) = 13.978
      U(I,3) = 1.426
      GAMMA(I,I) = 9.260
      GO TO 211
  404 ZN(I) = 2.03
      BETA(I,I) = -19.0/27.204
      U(I,1) = 43.182
      U(I,2) = 17.416
      U(I,3) = 1.954
      GAMMA(I,I) = 10.366
  211 CONTINUE
      DO 220 I = 1,NA
      DO 219 J = 1,NA
      IF((NAT(I).GT.10).OR.(NAT(J).GT.10)) GO TO 700
      BETA(I,J) = 0.5*(BETA(I,I)+BETA(J,J))
      GO TO 701
  700 BETA(I,J) = 0.5*(BETA(I,I)+BETA(J,J))*0.75
  701 CONTINUE
  219 CONTINUE
      WRITE(IY,105) I,XC(I),YC(I),ZC(I),NAT(I),U(I,1),U(I,2),U(I,3)
      U(I,1) = U(I,1) / 27.204
      U(I,2) = U(I,2)/ 27.204
  220 U(I,3) = U(I,3)/ 27.204
C     THE VALUE OF KAPPA IS 0.585
      AII(1)=0.585
      NELNS = -KHARGE
      IA = 1
      DO 301  I = 1,NA
      NFIRST(I) = IA
      NI = NAT(I)
      IF(NI.GT.2) GO TO 302
      NELNS = NELNS + NAT(I)
      IB = IA
      GO TO 303
  302 IF(NI.GT.10) GO TO 304
      NELNS = NELNS + NAT(I)-2
      IB = IA +3
      GO TO 303
  304 NELNS = NELNS + NAT(I)-10
```

```
          IB = IA + 8
      303 NLAST(I) = IB
      301 IA = IB+1
          N =NLAST(NA)
          NOCC = NELNS/2
          RETURN
       99 STOP
          END

          SUBROUTINE REPLSN
          COMMON /ONE/ N,NA,NOCC,ICHGE,IX,IY,I,J,KI,KJ,ISUB,IDUMB,ISTOP,
         1E,TCONV,MU,MO
          COMMON/SIX/ U(35,3),R(35,35),BETA(35,35),GAMMA(35,35)
    C     GAMMAS OBTAINED BY THE MATAGA METHOD.
          NAMII = NA - 1
          DO 104 I = 1,NAMII
          JJ = I + 1
          DO 104 J = JJ,NA
          D = 54.408/(GAMMA(I,I) + GAMMA(J,J))
          B = R(I,J)/0.529
          GAMMA(I,J) = (1.0/(B + D)) * 27.204
      104 GAMMA(J,I) = GAMMA(I,J)
          DO 105 I = 1,NA
          DO 105 J = 1,NA
      105 GAMMA(I,J) = GAMMA(I,J) / 27.204
          RETURN
          END

          SUBROUTINE BET1
          COMMON /ONE/ N,NA,NOCC,ICHGE,IX,IY,I,J,KI,KJ,ISUB,IDUMB,ISTOP,
         1E,TCONV,MU,MO
          COMMON/TWO/ AII(100),AIII(100),XC(35),YC(35),ZC(35),A(100,100),
         1            BETAO(100,100),C(100,100)
          COMMON/THREE/ NAT(35),S(15),ZN(35),ZCHG(35),Q(35)
          COMMON/FOUR/ P(35,35),PI(35,35)
          COMMON/FIVE/ PO(100,100),POI(100,100)
          COMMON/SIX/ U(35,3),R(35,35),BETA(35,35),GAMMA(35,35)
          COMMON/GEOD/ EA(3),T(9,9),TS(9,9),SA(20)
          DO 699 I = 1,N
          DO 699 J = 1,N
      699 BETAO(I,J) = 0.0
          KI = 1
          KJ = 1
          DO 717 I = 1,NA
          DO 714 J = 1,NA
          IF(I - J) 705, 701, 705
      701 IF(NAT(J) - 1) 703, 703, 704
      703 KJ = KJ + 1
          GO TO 714
      704 IF(NAT(J)-10) 801,801,802
      801 KJ = KJ+4
          GO TO 714
      802 KJ = KJ+9
          GO TO 714
      705 KD = NAT(I) + NAT(J)
          IF(KD - 2) 707, 706, 707
      706 CALL OVLP1
          GO TO 714
      707 IF(NAT(I) - 1) 709, 708, 709
```

```
  708 ICHGE = 2
      IF(NAT(J).GT.10) GO TO 803
      CALL OVLP2
      CALL GEOM1
      GO TO 714
  803 CONTINUE
      DO 1 IN=1,20
    1 SA(IN) = 0.0
      CALL OVLP5
      CALL RELVEC
      CALL MATS
      CALL HARMTR
      CALL GEOMD
      GO TO 714
  709 IF(NAT(J) - 1) 711, 710, 711
  710 ICHGE = 1
      IF(NAT(I).GT.10) GO TO 810
      CALL OVLP2
      CALL GEOM1
      GO TO 714
  810 CONTINUE
      DO 2 IN=1,20
    2 SA(IN) = 0.0
      CALL OVLP5
      CALL RELVEC
      CALL MATS
      CALL HARMTR
      CALL GEOMD
      GO TO 714
  711 IF(NAT(I) - NAT(J)) 713, 712, 713
  712 IF(NAT(I).GT.10) GO TO 804
      CALL OVLP3
      CALL GEOM2
      GO TO 714
  804 CONTINUE
      DO 3 IN=1,20
    3 SA(IN) = 0.0
      CALL OVLP8
      CALL RELVEC
      CALL MATS
      CALL HARMTR
      CALL GEOMD
      GO TO 714
  713 IF((NAT(J).GT.10).AND.(NAT(I).GT.10)) GO TO 805
      IF(NAT(I).GT.10) GO TO 806
      IF(NAT(J).GT.10) GO TO 807
      CALL OVLP4
      CALL GEOM2
      GO TO 714
  805 CONTINUE
      DO 4 IN=1,20
    4 SA(IN) = 0.0
      CALL OVLP7
      CALL RELVEC
      CALL MATS
      CALL HARMTR
      CALL GEOMD
      GO TO 714
  806 CONTINUE
```

```
          DO 5 IN=1,20
    5 SA(IN) = 0.0
      CALL OVLP6
      CALL RELVEC
      CALL MATS
      CALL HARMTR
      CALL GEOMD
      GO TO 714
  807 CONTINUE
      DO 6 IN=1,20
    6 SA(IN) = 0.0
      CALL OVLP6
      CALL RELVEC
      CALL MATS
      CALL HARMTR
      CALL GEOMD
      GO TO 714
  714 CONTINUE
      IF(NAT(I) - 1) 716, 715, 716
  715 KI = KI + 1
      KJ = 1
      GO TO 717
  716 IF(NAT(I).LT.10) GO TO 718
      KI = KI+9
      KJ = 1
      GO TO 717
  718 KI = KI+4
      KJ = 1
  717 CONTINUE
      AII(1) = 1.0
      RETURN
      END

      SUBROUTINE OVLP1
C     CALCULATION OF THE OVERLAP INTEGRAL BETWEEN TWO IS FUNCTIONS
      COMMON /ONE/ N,NA,NOCC,ICHGE,IX,IY,I,J,KI,KJ,ISUB,IDUMB,ISTOP,
     1E,TCONV,MU,MO
      COMMON/TWO/ AII(100),AIII(100),XC(35),YC(35),ZC(35),A(100,100),
     1            BETAO(100,100),C(100,100)
      COMMON/THREE/ NAT(35),S(15),ZN(35),ZCHG(35),Q(35)
      COMMON/FOUR/ P(35,35),PI(35,35)
      COMMON/FIVE/ PO(100,100),POI(100,100)
      COMMON/SIX/ U(35,3), R(35,35),BETA(35,35),GAMMA(35,35)
      DIMENSION AA(5)
      W = 2.26843 * R(I,J)
      AA(1) = EXP(-W)/W
      AA(3) = 2. * AA(1) * (.5 + 1./W + 1./(W**2))
      S(1) = ((W**3)/6.) * (3. *AA(3) - AA(1))
      BETAO(KI,KJ) = BETA(I,J) * S(1)
      KJ = KJ + 1
      RETURN
      END

      SUBROUTINE OVLP2
C     CALCULATION OF THE OVERLAP INTEGRAL BETWEEN 1S AND2S, 2P SIGMA
      COMMON /ONE/ N,NA,NOCC,ICHGE,IX,IY,I,J,KI,KJ,ISUB,IDUMB,ISTOP,
     1E,TCONV,MU,MO
      COMMON/THREE/ NAT(35),S(15),ZN(35),ZCHG(35),Q(35)
      COMMON/FOUR/ P(35,35),PI(35,35)
```

5 PROGRAM LISTINGS

```
      COMMON/FIVE/ PO(100,100),POI(100,100)
      COMMON/SIX/ U(35,3),R(35,35),BETA(35,35),GAMMA(35,35)
      DIMENSION AA(5),B(5)
      DIMENSION BAP(5),BAN(5)
      W = .5 *(ZN(I) + ZN(J)) * R(I,J)/.529
      T = ABS((ZN(I) - ZN(J))/(ZN(I) + ZN(J)))
      T = -T
      PT = W * T
      AA(1) = EXP(-W)/W
      DO 101 K = 2,4
      TK = K - 1
  101 AA(K) = (TK/W) * AA(K-1) + AA(1)
      BAP(1) = EXP(-PT)/PT
      BAN(1) = EXP(PT)/PT
      B(1) = BAN(1) - BAP(1)
      SIX = -1.
      DO 103 K = 2,5
      TK = K-1
      BAN(K) = (TK/PT)*BAN(K-1) + SIX* BAN(1)
      BAP(K) = (TK/PT) * BAP(K-1) + BAP(1)
  103 SIX = -SIX
      DO 122 K = 2,5
  122 B(K) = BAN(K) - BAP(K)
      S(1) = .07217 * W**4 *(1.+T)**1.5*(1.-T)**2.5* (AA(4) *B(1)-
     1AA(3) *B(2)- AA(2)* B(3) +AA(1) *B(4))
      S(2) = .125 * W**4 *(1. + T)**1.5 *(1.-T)**2.5 * (-AA(4) *B(2) +
     1AA(3) *B(1) + AA(2)*B(4) -AA(1)*B(3))
      RETURN
      END


      SUBROUTINE OVLP3
C     CALCULATION OF THE OVERLAP INTEGRALS BETWEEN C-C,N-N,O-O
      COMMON /ONE/ N,NA,NOCC,ICHGE,IX,IY,I,J,KI,KJ,ISUB,IDUMB,ISTOP,
     1E,TCONV,MU,MO
      COMMON/TWO/ AII(100),AIII(100),XC(35),YC(35),ZC(35),A(100,100),
     1            BETAO(100,100),C(100,100)
      COMMON/THREE/ NAT(35),S(15),ZN(35),ZCHG(35),Q(35)
      COMMON/FOUR/ P(35,35),PI(35,35)
      COMMON/FIVE/ PO(100,100),POI(100,100)
      COMMON/SIX/ U(35,3),R(35,35),BETA(35,35),GAMMA(35,35)
      DIMENSION AA(5)
      W = .5 *(ZN(I) + ZN(J)) * R(I,J)/.529
      AA(1) = EXP(-W)/W
      DO 101 K = 2,5
      TK = K - 1
  101 AA(K) = (TK/W) * AA(K-1) + AA(1)
      S(1) = .00277778 * W**5 * (15. *AA(5) -10. * AA(3) + 3. * AA(1))
      S(2) = .0096227 * W **5 *(5.*AA(4) - AA(2))
      S(3) = .00833333 *W**5 *(5.*AA(5) -18. *AA(3) + 5. * AA(1))
      S(3) = -S(3)
      S(4) = .00833333 *W**5 *(5.*AA(5) - 6. *AA(3) + AA(1))
      KD=NAT(I)+NAT(J)
      IF (KD-12) 100,102,100
  102 S(4)=S(4)*0.570
      GO TO 103
  100 S(4)=S(4)*AII(1)
  103 S(5)=S(2)
      RETURN
      END
```

```
      SUBROUTINE OVLP4
C     CALCULATION OF S BETWEEN TWO DIFFERENT SECOND-ROW ELEMENTS
      COMMON /ONE/ N,NA,NOCC,ICHGE,IX,IY,I,J,KI,KJ,ISUB,IDUMB,ISTOP,
     1E,TCONV,MU,MO
      COMMON/TWO/ AII(100),AIII(100),XC(35),YC(35),ZC(35),A(100,100),
     1            BETAO(100,100),C(100,100)
      COMMON/THREE/ NAT(35),S(15),ZN(35),ZCHG(35),Q(35)
      COMMON/FOUR/ P(35,35),PI(35,35)
      COMMON/FIVE/ PO(100,100),POI(100,100)
      COMMON/SIX/ U(35,3),R(35,35),BETA(35,35),GAMMA(35,35)
      DIMENSION AA(5),B(5)
      DIMENSION BAP(5),BAN(5)
      W = .5 *(ZN(I) + ZN(J)) * R(I,J)/.529
      T = ABS((ZN(I) - ZN(J))/(ZN(I) + ZN(J)))
      PT=W*T
      AA(1) = EXP(-W)/W
      DO 101 K = 2,5
      TK = K - 1
  101 AA(K) = (TK/W) * AA(K-1) + AA(1)
      BAP(1) = EXP(-PT)/PT
      BAN(1) = EXP(PT)/PT
      B(1) = BAN(1) - BAP(1)
      SIX = -1.
      DO 103 K = 2,5
      TK = K - 1
      BAN(K) = (TK/PT) * BAN(K-1) + SIX * BAN(1)
      BAP(K) = (TK/PT) * BAP(K-1) + BAP(1)
  103 SIX = -SIX
      DO 122 K = 2,5
  122 B(K) = BAN(K) - BAP(K)
      S(1) = .02083 *W**5 * (1.-T**2) **2.5 * (AA(5)*B(1) - 2.*AA(3)*
     1B(3) + AA(1) *B(5))
      S(2) = .036085*W**5*(1.-T**2)**2.5*(AA(4)*(B(1)-B(3))+AA(2)*
     1(B(5)-B(3))+B(2)*(AA(3) - AA(5)) + B(4)*(AA(3)-AA(1)))
      S(3) = .0625*W**5*(1.-T**2)**2.5*(B(3) *(AA(1) + AA(5)) - AA(3)*
     1(B(1) + B(5)))
      S(3) = -S(3)
      S(4) = .03125 *W**5*(1.-T**2)**2.5*(AA(5)*(B(1)-B(3))+ AA(3) *
     1(B(5) - B(1)) + AA(1) * (B(3) - B(5)))
      KD=NAT(I)+NAT(J)
      IF (KD-14) 300,301,300
  301 S(4)=S(4)*0.585
      GO TO 302
  300 S(4)=S(4)*AII(1)
  302 T=-T
      PT = -PT
      BAP(1) = EXP(-PT)/PT
      BAN(1) = EXP(PT)/PT
      B(1) = BAN(1) - BAP(1)
      SIX = -1.
      DO 203 K = 2,5
      TK = K - 1
      BAN(K) = (TK/PT) * BAN(K-1) + SIX * BAN(1)
      BAP(K) = (TK/PT) * BAP(K-1) + BAP(1)
  203 SIX = -SIX
      DO 222 K = 2,5
  222 B(K) = BAN(K) - BAP(K)
      S(5) = .036085*W**5*(1.-T**2)**2.5 *(AA(4)*(B(1) - B(3)) + AA(2)*
     1(B(5)-B(3)) + B(2) *(AA(3) - AA(5)) + B(4) * (AA(3) - AA(1)))
```

```
      RETURN
      END

      SUBROUTINE OVLP5
C     CALCULATION OF S BETWEEN H AND THIRD-ROW ELEMENTS
      COMMON/ONE/ N,NA,NOCC,ICHGE,IX,IY,I,J,KI,KJ,ISUB,IDUMB,ISTOP,
     1            E,TCONV,MU,MO
      COMMON/TWO/ AII(100),AIII(100),XC(35),YC(35),ZC(35),A(100,100),
     1            BETAO(100,100),C(100,100)
      COMMON/THREE/ NAT(35),S(15),ZN(35),ZCHG(35),Q(35)
      COMMON/FOUR/ P(35,35),PI(35,35)
      COMMON/FIVE/ PO(100,100),POI(100,100)
      COMMON/SIX/ U(35,3),R(35,35),BETA(35,35),GAMMA(35,35)
      COMMON/GEOD/ EE(3),TA(9,9),TS(9,9),SA(20)
      DIMENSION AA(5),B(5)
      DIMENSION BAP(5),BAN(5)
      W = 0.5*(ZN(I)+ZN(J))*R(I,J)/0.529
      T = ABS((ZN(I)-ZN(J))/(ZN(I)+ZN(J)))
      PT = W*T
      AA(1) = EXP(-W)/W
      DO 101 K=2,5
      TK = K-1
  101 AA(K) = (TK/W)*AA(K-1) + AA(1)
      BAP(1) = EXP(-PT)/PT
      BAN(1) = EXP(+PT)/PT
      B(1) = BAN(1) - BAP(1)
      SIX = -1.0
      DO 103 K=2,5
      TK = K-1
      BAN(K) = (TK/PT)*BAN(K-1) + SIX*BAN(1)
      BAP(K) = (TK/PT)*BAP(K-1) + BAP(1)
  103 SIX = -SIX
      DO 122 K=2,5
  122 B(K) = BAN(K) - BAP(K)
      S(1) = 0.013176*W**5.0*(1.0+T)**1.5*(1.0-T)**3.5
     1      *(AA(5)*B(1) - 2.0*AA(4)*B(2) +2.0*AA(2)*B(4) - AA(1)*B(5))
      S(2) = 0.022822*W**5.0*(1.0+T)**1.5*(1.0-T)**3.5
     1      *(AA(4)*(B(1)+B(3)) - AA(2)*(B(3)+B(5))
     2       - B(2)*(AA(3)+AA(5)) + B(4)*(AA(1)+AA(3)))
      S(3) = 0.014731*W**5.0*(1.0+T)**1.5*(1.0-T)**3.5
     1      *((AA(1)*B(5)-AA(5)*B(1)) + 4.0*(AA(2)*B(4)-AA(4)*B(2))
     2       + 3.0*(AA(3)*B(1)-AA(1)*B(3)) + 3.0*(AA(5)*B(3)-AA(3)*B(5)))
      SA(1) = S(1)
      SA(8) = S(2)
      SA(11) = S(3)*0.300
      IF(NAT(I).LT.10) GO TO 200
      SA(8) = 0.0
      SA(11) = 0.0
      SA(2) = S(2)
      SA(3) = S(3)*0.300
  200 CONTINUE
      RETURN
      END
```

```
      SUBROUTINE OVLP6
C     CALCULATION OF S BETWEEN SECOND AND THIRD ROW ELEMENTS
      COMMON/ONE/ N,NA,NOCC,ICHGE,IX,IY,I,J,KI,KJ,ISUB,IDUMB,ISTOP,
     1            E,TCONV,MU,MO
      COMMON/TWO/ AII(100),AIII(100),XC(35),YC(35),ZC(35),A(100,100),
     1            BETAO(100,100),C(100,100)
      COMMON/THREE/ NAT(35),S(15),ZN(35),ZCHG(35),Q(35)
      COMMON/FOUR/ P(35,35),PI(35,35)
      COMMON/FIVE/ PO(100,100),POI(100,100)
      COMMON/SIX/ U(35,3),R(35,35),BETA(35,35),GAMMA(35,35)
      COMMON/GEOD/ EE(3),TA(9,9),TS(9,9),SA(20)
      DIMENSION AA(7),B(7),BAP(7),BAN(7)
      W = 0.5*(ZN(I)+ZN(J))*R(I,J)/0.529
      T = ABS((ZN(I)-ZN(J))/(ZN(I)+ZN(J)))
      PT = W*T
      AA(1) = EXP(-W)/W
      DO 101 K=2,7
      TK = K-1
  101 AA(K) = (TK/W)*AA(K-1) + AA(1)
      BAP(1) = EXP(-PT)/PT
      BAN(1) = EXP(+PT)/PT
      B(1) = BAN(1) - BAP(1)
      SIX = -1.0
      DO 103 K=2,7
      TK = K-1
      BAN(K) = (TK/PT)*BAN(K-1) + SIX*BAN(1)
      BAP(K) = (TK/PT)*BAP(K-1) + BAP(1)
  103 SIX = -SIX
      DO 122 K=1,7
  122 B(K) = BAN(K) - BAP(K)
      S(1) = 0.003804*W**6.0*(1.0+T)**2.5*(1.0-T)**3.5
     1      *(AA(6)*B(1) - AA(5)*B(2) - 2.0*AA(4)*B(3)
     2      + 2.0*AA(3)*B(4) + AA(2)*B(5) - AA(1)*B(6))
      S(2) = 0.006588*W**6.0*(1.0+T)**2.5*(1.0-T)**3.5
     1      *(AA(1)*B(5) - AA(2)*B(6) - 2.0*AA(3)*B(3)
     2      + 2.0*AA(4)*B(4) + AA(5)*B(1) - AA(6)*B(2))
      S(3) = 0.006588*W**6.0*(1.0+T)**2.5*(1.0-T)**3.5
     1      *(AA(5)*(B(1)-2.0*B(3)) + AA(2)*(2.0*B(4)-B(6))
     2      + B(2)*(AA(6)-2.0*AA(4)) + B(5)*(2.0*AA(3)-AA(1)))
      S(4) = 0.011411*W**6.0*(1.0+T)**2.5*(1.0-T)**3.5
     1      *(-AA(3)*(B(2)+B(6))  + AA(4)*(B(1)+B(5))
     2      + B(4)*(AA(1)+AA(5)) - B(3)*(AA(2)+AA(6)))
      S(5) = 0.005705*W**6.0*(1.0+T)**2.5*(1.0-T)**3.5
     1      *(AA(6)*(B(1)-B(3)) + AA(5)*(B(4)-B(2))
     2      + AA(4)*(B(5)-B(1)) + AA(3)*(B(2)-B(6))
     3      + AA(2)*(B(3)-B(5)) + AA(1)*(B(6)-B(4)))
      S(6) = 0.012758*W**6.0*(1.0+T)**2.5*(1.0-T)**3.5
     1      *(- AA(6)*(B(2)-B(4)) + AA(5)*(B(1)-B(3))
     2      + AA(4)*(B(2)-B(6)) - AA(3)*(B(1)-B(5))
     3      - AA(2)*(B(4)-B(6)) + AA(1)*(B(3)-B(5)))
      S(7) = 0.004253*W**6.0*(1.0+T)**2.5*(1.0-T)**3.5
     1      *(- AA(1)*(3.0*B(4)-B(6)) - AA(2)*(3.0*B(3)-5.0*B(5))
     2      + AA(3)*(3.0*B(2)+4.0*B(4)-3.0*B(6))
     3      + AA(4)*(3.0*B(1)-4.0*B(3)-3.0*B(5))
     4      - AA(5)*(5.0*B(2)-3.0*B(4)) - AA(6)*(B(1)-3.0*B(3)))
      S(8) = 0.007366*W**6.0*(1.0+T)**2.5*(1.0-T)**3.5
     1      *(- AA(1)*(3.0*B(3)-B(5)) + AA(2)*(B(4)+B(6))
     2      + AA(3)*(3.0*B(1)+B(5)) - AA(4)*(B(2)+3.0*B(6))
     3      - AA(5)*(B(1)+B(3)) - AA(6)*(B(2)-3.0*B(4)))
```

```
      SA(1) = S(1)
      SA(2) = S(3)
      SA(4) = S(5)*0.585
      SA(6) = S(5)*0.585
      SA(8) = S(2)
      SA(9) = S(4)
      SA(11) = S(7)*0.300
      SA(12) = S(8)*0.300
      SA(14) = S(6)*0.300
      SA(16) = S(6)*0.300
      IF(NAT(I).LT.10) GO TO 700
      SA(11) = 0.0
      SA(12) = 0.0
      SA(14) = 0.0
      SA(16) = 0.0
      SA(3) = S(7)*0.300
      SA(5) = S(6)*0.300
      SA(7) = S(6)*0.300
      SA(10) = S(8)*0.300
  700 CONTINUE
      RETURN
      END

      SUBROUTINE OVLP7
C     CALCULATION OF S BETWEEN DIFFERENT THIRD ROW ATOMS
      COMMON/ONE/ N,NA,NOCC,ICHGE,IX,IY,I,J,KI,KJ,ISUB,IDUMB,ISTOP,
     1            E,TCONV,MU,MO
      COMMON/TWO/ AII(100),AIII(100),XC(35),YC(35),ZC(35),A(100,100),
     1            BETAO(100,100),C(100,100)
      COMMON/THREE/ NAT(35),S(15),ZN(35),ZCHG(35),Q(35)
      COMMON/FOUR/ P(35,35),PI(35,35)
      COMMON/FIVE/ PO(100,100),POI(100,100)
      COMMON/SIX/ U(35,3),R(35,35),BETA(35,35),GAMMA(35,35)
      COMMON/GEOD/ EE(3),TA(9,9),TS(9,9),SA(20)
      DIMENSION AA(8),B(8),BAP(8),BAN(8)
      W = 0.5*(ZN(I)+ZN(J))*R(I,J)/0.529
      T = ABS((ZN(I)-ZN(J))/(ZN(I)+ZN(J)))
      PT = W*T
      AA(1) = EXP(-W)/W
      DO 101 K=2,8
      TK = K-1
  101 AA(K) = (TK/W)*AA(K-1) + AA(1)
      BAP(1) = EXP(-PT)/PT
      BAN(1) = EXP(+PT)/PT
      B(1) = BAN(1) - BAP(1)
      SIX = -1.0
      DO 103 K=2,8
      TK = K-1
      BAN(K) = (TK/PT)*BAN(K-1) + SIX*BAN(1)
      BAP(K) = (TK/PT)*BAP(K-1) + BAP(1)
  103 SIX = -SIX
      DO 122 K=2,8
  122 B(K) = BAN(K) - BAP(K)
      S(1) = 0.000694*W**7.0*(1.0+T)**3.5*(1.0-T)**3.5
     1       *(AA(7)*B(1) - 3.0*AA(5)*B(3) + 3.0*AA(3)*B(5)
     2        - AA(1)*B(7))
      S(2) = 0.001203*W**7.0*(1.0+T)**3.5*(1.0-T)**3.5
     1       *(- AA(7)*B(2) + AA(6)*(B(1)-B(3))
     2         + AA(5)*(B(2)+2.0*B(4)) + 2.0*AA(4)*(B(5)-B(3))
```

THEORY OF NMR PARAMETERS

```
     3             - AA(3)*(2.0*B(4)+B(6)) + AA(2)*(B(5)-B(7)) + AA(1)*B(6))
      S(3) = 0.002083*W**7.0*(1.0+T)**3.5*(1.0-T)**3.5
     1           *( AA(1)*B(5) - AA(3)*(B(7)+2.0*B(3))
     2               + AA(5)*(B(1)+2.0*B(5)) - AA(7)*B(3))
      S(4) = 0.001042*W**7.0*(1.0+T)**3.5*(1.0-T)**3.5
     1           *( AA(7)*(B(1)-B(3)) + AA(5)*(2.0*B(5)-B(1)-B(3))
     2               + AA(3)*(2.0*B(3)-B(5)-B(7)) + AA(1)*(B(7)-B(5)))
      S(5) = 0.002329*W**7.0*(1.0+T)**3.5*(1.0-T)**3.5
     1           *( AA(1)*(B(4)-B(6)) + AA(2)*(B(3)-2.0*B(5)+B(7))
     2               - AA(3)*(B(2)+B(4)-2.0*B(6))
     3               - AA(4)*(B(1)-B(3)-B(5)-B(7))
     4               + AA(5)*(2.0*B(2)-B(4)-B(6))
     5               + AA(6)*(B(1)-2.0*B(3)+B(5)) - AA(7)*(B(2)-B(4)))
      S(6) = 0.005208*W**7*(1.0+T)**3.5*(1.0-T)**3.5
     1           *((AA(1)-AA(7))*(B(3)-B(5)) - (AA(3)-AA(5))*(B(1)-B(7)))
      S(9) = 0.000776*W**7*(1.0+T)**3.5*(1.0-T)**3.5
     1           *( - AA(1)*(3.0*B(5)-B(7)) - 6.0*AA(2)*(B(4)-B(6))
     2               + 3.0*AA(3)*(3.0*B(5)-B(7)) + 6.0*AA(4)*(B(2)-B(6))
     3               + 3.0*AA(5)*(B(1)-3.0*B(3)) - 6.0*AA(6)*(B(2)-B(4))
     4               - AA(7)*(B(1)-3.0*B(3)))
      S(10) = 0.001345*W**7.0*(1.0+T)**3.5*(1.0-T)**3.5
     1           *( - AA(1)*(3.0*B(4)-B(6))
     2                - AA(2)*(3.0*B(3)-2.0*B(5)-B(7))
     3                + AA(3)*(3.0*B(2)+B(4)+2.0*B(6))
     4                + AA(4)*(3.0*B(1)-B(3)+B(5)-3.0*B(7))
     5                - AA(5)*(2.0*B(2)+B(4)+3.0*B(6))
     6                - AA(6)*(B(1)+2.0*B(3)-3.0*B(5))
     7                - AA(7)*(B(2)-3.0*B(4)))
      S(11) = 0.000869*W**7.0*(1.0+T)**3.5*(1.0-T)**3.5
     1           *( - AA(1)*(9.0*B(3)-6.0*B(5)+B(7))
     2                + 3.0*AA(3)*(3.0*B(1)-B(5)+2.0*B(7))
     3                - 3.0*AA(5)*(2.0*B(1)-B(3)+3.0*B(7))
     4                + AA(7)*(B(1)-6.0*B(3)+9.0*B(5)))
      S(12) = 0.001302*W**7*(1.0+T)**3.5*(1.0-T)**3.5
     1           *( - AA(1)*(B(3)-2.0*B(5)+B(7))
     2                + AA(3)*(B(1)-3.0*B(5)+2.0*B(7))
     3                - AA(5)*(2.0*B(1)-3.0*B(3)+B(7))
     4                + AA(7)*(B(1)-2.0*B(3)+B(5)))
      T = -T
      PT = -PT
      BAP(1) = EXP(-PT)/PT
      BAN(1) = EXP(+PT)/PT
      B(1) = BAN(1) - BAP(1)
      SIX = -1.0
      DO 203 K=2,8
      TK = K-1
      BAN(K) = (TK/PT)*BAN(K-1) + SIX*BAN(1)
      BAP(K) = (TK/PT)*BAP(K-1) + BAP(1)
  203 SIX = -SIX
      DO 222 K=2,8
  222 B(K) = BAN(K) - BAP(K)
      S(7) = 0.001203*W**7.0*(1.0+T)**3.5*(1.0-T)**3.5
     1           *( - AA(7)*B(2) + AA(6)*(B(1)-B(3))
     2                + AA(5)*(B(2)+2.0*B(4)) + 2.0*AA(4)*(B(5)-B(3))
     3                - AA(3)*(2.0*B(4)+B(6)) + AA(2)*(B(5)-B(7)) + AA(1)*B(6))
      S(8) = 0.002329*W**7.0*(1.0+T)**3.5*(1.0-T)**3.5
     1           *( AA(1)*(B(4)-B(6)) + AA(2)*(B(3)-2.0*B(5)+B(7))
     2                - AA(3)*(B(2)+B(4)-2.0*B(6))
     3                - AA(4)*(B(1)-B(3)-B(5)+B(7))
```

```
    4            + AA(5)*(2.0*B(2)-B(4)-B(6))
    5            + AA(6)*(B(1)-2.0*B(3)+B(5)) - AA(7)*(B(2)-B(4)))
      S(13) = 0.000776*W**7.0*(1.0+T)**3.5*(1.0-T)**3.5
    1        *( - AA(1)*(3.0*B(5)-B(7)) - 6.0*AA(2)*(B(4)-B(6))
    2           + 3.0*AA(3)*(3.0*B(5)-B(7)) + 6.0*AA(4)*(B(2)-B(6))
    3           + 3.0*AA(5)*(B(1)-3.0*B(3)) - 6.0*AA(6)*(B(2)-B(4))
    4           -AA(7)*(B(1)-3.0*B(3)))
      S(14) = 0.001345*W**7.0*(1.0+T)**3.5*(1.0-T)**3.5
    1        *( - AA(1)*(3.0*B(4)-B(6))
    2           - AA(2)*(3.0*B(3)-2.0*B(5)-B(7))
    3           + AA(3)*(3.0*B(2)+B(4)+2.0*B(6))
    4           + AA(4)*(3.0*B(1)-B(3)+B(5)-3.0*B(7))
    5           - AA(5)*(2.0*B(2)+B(4)+3.0*B(6))
    6           - AA(6)*(B(1)+2.0*B(3)-3.0*B(5))
    7           - AA(7)*(B(2)-3.0*B(4)))
      SA(1) = S(1)
      SA(2) = S(7)
      SA(3) = S(13)*0.300
      SA(4) = S(4)*0.585
      SA(5) = S(8)*0.300
      SA(6) = S(4)*0.585
      SA(7) = S(8)*0.300
      SA(8) = S(2)
      SA(9) = S(3)
      SA(10) = S(14)*0.300
      SA(11) = S(9)*0.300
      SA(12) = S(10)*0.300
      SA(13) = S(11)*0.300
      SA(14) = S(5)*0.300
      SA(15) = S(6)*0.300
      SA(16) = S(5)*0.300
      SA(17) = S(6)*0.300
      SA(18) = S(12)*0.300
      SA(19) = S(12)*0.300
      RETURN
      END

      SUBROUTINE OVLP8
C     CALCULATION OF S BETWEEN SAME ELEMENT OF THIRD ROW ATOMS
      COMMON/ONE/ N,NA,NOCC,ICHGE,IX,IY,I,J,KI,KJ,ISUB,IDUMB,ISTOP,
    1             E,TCONV,MU,MO
      COMMON/TWO/ AII(100),AIII(100),XC(35),YC(35),ZC(35),A(100,100),
    1             BETAO(100,100),C(100,100)
      COMMON/THREE/ NAT(35),S(15),ZN(35),ZCHG(35),Q(35)
      COMMON/FOUR/ P(35,35),PI(35,35)
      COMMON/FIVE/ PO(100,100),POI(100,100)
      COMMON/SIX/ U(35,3),R(35,35),BETA(35,35),GAMMA(35,35)
      COMMON/GEOD/ EE(3),TA(9,9),TS(9,9),SA(20)
      DIMENSION AA(8)
      W = 0.5*(ZN(I)+ZN(J))*R(I,J)/0.529
      AA(1) = EXP(-W)/W
      DO 102 K=2,8
      TK = K-1
  102 AA(K) = (TK/W)*AA(K-1) + AA(1)
      S(1) = 0.000040*W**7.0
    1      *(35.0*AA(7) - 35.0*AA(5) + 21.0*AA(3) - 5.0*AA(1))
      S(2) = 0.000046*W**7.0
    1      *(35.0*AA(6) - 14.0*AA(4) + 3.0*AA(2))
      S(3) = 0.000040*W**7.0
```

```
      1         *( 21.0*AA(1) - 85.0*AA(3) + 147.0*AA(5) - 35.0*AA(7))
      S(4) = 0.000040*W**7.0
      1         *( 35.0*AA(7) - 49.0*AA(5) + 17.0*AA(3) - 3.0*AA(1))
      S(5) = 0.000355*W**7.0
      1         *( AA(2) - 8.0*AA(4) + 7.0*AA(6))
      S(6) = 0.020833*W**7.0
      1         *((AA(1)-AA(7))/15.0 - 3.0*(AA(3)-AA(5))/7.0)
      S(7) = S(2)
      S(8) = S(5)
      S(9) = 0.000710*W**7.0
      1         *( 3.0*AA(3) - AA(1))
      S(10) = 0.000410*W**7.0
      1         *( - 3.0*AA(2) + 16.0*AA(4) - 7.0*AA(6))
      S(11) = 0.000198*W**7.0
      1         *( - 17.0*AA(1) + 81.0*AA(3) - 55.0*AA(5) + 7.0*AA(7))
      S(12) = 0.000198*W**7.0
      1         *( - AA(1) + 9.0*AA(3) - 15.0*AA(5) + 7.0*AA(7))
      SA(1) = S(1)
      SA(2) = S(7)
      SA(3) = S(9)*0.300
      SA(4) = S(4)*0.585
      SA(5) = S(8)*0.300
      SA(6) = S(4)*0.585
      SA(7) = S(8)*0.300
      SA(8) = S(2)*0.585
      SA(9) = S(3)
      SA(10) = S(10)*0.300
      SA(11) = S(9)*0.300
      SA(12) = S(10)*0.300
      SA(13) = S(11)*0.300
      SA(14) = S(5)*0.300
      SA(15) = S(6)*0.300
      SA(16) = S(5)*0.300
      SA(17) = S(6)*0.300
      SA(18) = S(12)*0.300
      SA(19) = S(12)*0.300
      RETURN
      END
      
      SUBROUTINE GEOM1
C     MODIFICATION OF BETAS BY OVERLAP AND GEOMETRY H WITH 2ND ROW
      COMMON /ONE/ N,NA,NOCC,ICHGE,IX,IY,I,J,KI,KJ,ISUB,IDUMB,ISTOP,
     1E,TCONV,MU,MO
      COMMON/TWO/ AII(100),AIII(100),XC(35),YC(35),ZC(35),A(100,100),
     1            BETAO(100,100),C(100,100)
      COMMON/THREE/ NAT(35),S(15),ZN(35),ZCHG(35),Q(35)
      COMMON/FOUR/ P(35,35),PI(35,35)
      COMMON/FIVE/PO(100,100),POI(100,100)
      COMMON/SIX/ U(35,3),R(35,35),BETA(35,35),GAMMA(35,35)
      BETS = S(2) * BETA(I,J)
      XQ = XC(I) - XC(J)
      YQ = YC(I) - YC(J)
      ZQ = ZC(I) - ZC(J)
      BETAO(KI,KJ) = S(1) * BETA(I,J)
      B2 = BETS * XQ / R(I,J)
      B3 = BETS * YQ / R(I,J)
      B4 = BETS * ZQ / R(I,J)
      IF(ICHGE .EQ. 1) GO TO 111
      BETAO(KI,KJ+1) = B2
```

```
      BETAO(KI,KJ+2) = B3
      BETAO(KI,KJ+3) = B4
      KJ = KJ + 4
      GO TO 113
  111 BETAO(KI+1,KJ) = -B2
      BETAO(KI+2,KJ) = -B3
      BETAO(KI+3,KJ) = -B4
      KJ = KJ + 1
  113 RETURN
      END

      SUBROUTINE GEOM2
C     MODIFICATION OF BETAS BY S AND GEOMETRY SECOND ROW ELEMENTS
      COMMON /ONE/ N,NA,NOCC,ICHGE,IX,IY,I,J,KI,KJ,ISUB,IDUMB,ISTOP,
     1E,TCONV,MU,MO
      COMMON/TWO/ AII(100),AIII(100),XC(35),YC(35),ZC(35),A(100,100),
     1            BETAO(100,100),C(100,100)
      COMMON/THREE/ NAT(35),S(15),ZN(35),ZCHG(35),Q(35)
      COMMON/FOUR/ P(35,35),PI(35,35)
      COMMON/FIVE/ PO(100,100),POI(100,100)
      COMMON/SIX/ U(35,3),R(35,35),BETA(35,35),GAMMA(35,35)
      DIMENSION ARX(3)
      BETS = BETA(I,J)
      LI = KI
      KO = KJ
      BETAO(LI,KO) = S(1) * BETS
      XQ = XC(I) - XC(J)
      YQ = YC(I) - YC(J)
      ZQ = ZC(I) - ZC(J)
      ARX(1) = XQ
      ARX(2) = YQ
      ARX(3) = ZQ
      DO 100 IND = 1,3
      LINT = LI + IND
      KINT = KO + IND
  100 BETAO(LINT,KINT) = (-1.) * S(3) * BETS *(ARX(IND)**2)/(R(I,J)**2)
     1+S(4) * BETS * (1.-ARX(IND)**2/(R(I,J)**2))
      IF(NAT(I) - NAT(J)) 102,101,101
  101 S1 = S(2)
      S2 = S(5)
      GO TO 103
  102 S1 = S(5)
      S2 = S(2)
  103 BETAO(LI,KO+1) = S1 * BETS * XQ/R(I,J)
      BETAO(LI+1,KO) = S2 * BETS * (-1.) * XQ/R(I,J)
      BETAO(LI,KO + 2) = S1 * BETS * YQ/R(I,J)
      BETAO(LI+2,KO) = S2 * BETS *(-1.)*YQ/R(I,J)
      BETAO(LI,KO+3) = S1 *BETS * ZQ/R(I,J)
      BETAO(LI+3,KO) = S2 * BETS * (-1.) * ZQ/R(I,J)
      DO 200 IND = 1,2
      JD = IND + 1
      LINT = LI + IND
      KENT = KO + IND
      DO 200 JND = JD,3
      KINT = KO + JND
      LENT = LI + JND
      IF(ARX(IND)) 113,130,114
  113 IF(ARX(JND)) 120,130,119
  114 IF(ARX(JND)) 119,130,120
```

```
  119 SIX = 1.
      GO TO 131
  120 SIX = -1.
      GO TO 131
  130 BETAO(LINT,KINT) = 0.0
      BETAO(LENT,KENT) = 0.0
      GO TO 200
  131 BETAO(LINT,KINT) =   SIX *(ABS(ARX(IND) * ARX(JND)) /(R(I,J)**2)
     1*S(3) *BETS + ABS(ARX(IND)*ARX(JND))/(R(I,J)**2)*S(4) *BETS)
      BETAO(LENT,KENT) = BETAO(LINT,KINT)
  200 CONTINUE
      KJ = KO + 4
      RETURN
      END

      SUBROUTINE RELVEC
      COMMON/GEOD/ EA(3),T(9,9),TS(9,9),SA(20)
      COMMON/ONE/  N,NA,NOCC,ICHGE,IX,IY,II,JJ,KI,KJ,ISUB,IDUMB,ISTOP,
     1             E,TCONV,MU,MO
      COMMON/TWO/  AII(100),AIII(100),XC(35),YC(35),ZC(35),A(100,100),
     1             BETAO(100,100),C(100,100)
      COMMON/SIX/  U(35,3),R(35,35),BETA(35,35),GAMMA(35,35)
      D = R(II,JJ)
      EA(1) = (XC(JJ)-XC(II))/D
      EA(2) = (YC(JJ)-YC(II))/D
      EA(3) = (ZC(JJ)-ZC(II))/D
      RETURN
      END

      SUBROUTINE MATS
      COMMON/GEOD/ E(3),T(9,9),TS(9,9),SA(20)
      DO 1 I=1,9
      DO 1 J=1,9
      TS(I,J) = 0.0
    1 CONTINUE
      TS(1,1) = SA(1)
      TS(1,4) = SA(8)
      TS(2,2) = SA(4)
      TS(2,6) = SA(14)
      TS(3,3) = SA(6)
      TS(3,7) = SA(16)
      TS(4,1) = SA(2)
      TS(4,4) = SA(9)
      TS(4,5) = SA(12)
      TS(5,1) = SA(3)
      TS(5,4) = SA(10)
      TS(5,5) = SA(13)
      TS(1,5) = SA(11)
      TS(6,2) = SA(5)
      TS(6,6) = SA(15)
      TS(7,3) = SA(7)
      TS(7,7) = SA(17)
      TS(8,8) = SA(18)
      TS(9,9) = SA(19)
      RETURN
      END
```

5 PROGRAM LISTINGS

```
      SUBROUTINE HARMTR
      COMMON/GEOD/ E(3),T(9,9),TS(9,9),SA(20)
      MAXL = 2
      COST = E(3)
      IF((1.0-COST**2).GT.0.000000001) GO TO 20
   10 SINT = 0.0
      GO TO 30
   20 SINT = SQRT(1.0-COST**2)
   30 CONTINUE
      IF(SINT.GT.0.000000001) GO TO 50
   40 COSP = 1.0
      SINP = 0.0
      GO TO 70
   50 COSP = E(1)/SINT
   60 SINP = E(2)/SINT
   70 CONTINUE
      DO 80 I=1,9
      DO 80 J=1,9
   80 T(I,J) = 0.0
      T(1,1) = 1.0
      IF(MAXL.GT.1) GO TO 100
   90 IF(MAXL.GT.0) GO TO 110
      GO TO 120
  100 COS2T = COST**2 - SINT**2
      SIN2T = 2.0*SINT*COST
      COS2P = COSP**2 - SINP**2
      SIN2P = 2.0*SINP*COSP
C     TRANSFORMATION ELEMENTS FOR D FUNCTIONS
      SQRT3 = SQRT(3.0)
      T(5,5) = (3.0*COST**2 -1.0)/2.0
      T(5,6) = -SQRT3*SIN2T/2.0
      T(5,8) = SQRT3*SINT**2/2.0
      T(6,5) = SQRT3*SIN2T*COSP/2.0
      T(6,6) = COS2T*COSP
      T(6,7) = -COST*SINP
      T(6,8) = -T(6,5)/SQRT3
      T(6,9) = SINT*SINP
      T(7,5) = SQRT3*SIN2T*SINP/2.0
      T(7,6) = COS2T*SINP
      T(7,7) = COST*COSP
      T(7,8) = -T(7,5)/SQRT3
      T(7,9) = -SINT*COSP
      T(8,5) = SQRT3*SINT**2*COS2P/2.0
      T(8,6) = SIN2T*COS2P/2.0
      T(8,7) = -SINT*SIN2P
      T(8,8) = (1.0 + COST**2)*COS2P/2.0
      T(8,9) = -COST*SIN2P
      T(9,5) = SQRT3*SINT**2*SIN2P/2.0
      T(9,6) = SIN2T*SIN2P/2.0
      T(9,7) = SINT*COS2P
      T(9,8) = (1.0 + COST**2)*SIN2P/2.0
      T(9,9) = COST*COS2P
  110 CONTINUE
C     TRANSFORMATION MATRIX ELEMENTS FOR P FUNCTIONS
      T(2,2) = COST*COSP
      T(2,3) = -SINP
      T(2,4) = SINT*COSP
      T(3,2) = COST*SINP
      T(3,3) = COSP
```

```
      T(3,4) = SINT*SINP
      T(4,2) = -SINT
      T(4,4) = COST
  120 CONTINUE
      RETURN
      END

      SUBROUTINE GEOMD
      COMMON/GEOD/ EA(3),T(9,9),TS(9,9),SA(20)
      COMMON /ONE/ N,NA,NOCC,ICHGE,IX,IY,II,JJ,KI,KJ,ISUB,IDUMB,ISTOP,
     1             E,TCONV,MU,MO
      COMMON/TWO/ AII(100),AIII(100),XC(35),YC(35),ZC(35),A(100,100),
     1             BETAO(100,100),C(100,100)
      COMMON/THREE/ NAT(35),S(15),ZN(35),ZCHG(35),Q(35)
      COMMON/SIX/ U(35,3),R(35,35),BETA(35,35),GAMMA(35,35)
      DIMENSION INOR(35),TEMP(9,9)
      DO 1 IO=1,NA
      IF(NAT(IO).GT.10) GO TO 2
      IF(NAT(IO).GT.2)  GO TO 3
      INOR(IO) = 1
      GO TO 1
    3 INOR(IO) = 4
      GO TO 1
    2 INOR(IO) = 9
    1 CONTINUE
      NORBK = INOR(II)
      NORBL = INOR(JJ)
C     ROTATE INTEGRALS FROM DIATOMIC BASIS TO MOLECULAR BASIS
      DO 230 I=1,NORBK
      DO 230 J=1,NORBL
      TEMP(I,J) = 0.0
      DO 220 KK=1,NORBL
      TEMP(I,J) = TEMP(I,J) + T(J,KK)*TS(I,KK)
  220 CONTINUE
  230 CONTINUE
      DO 240 I=1,NORBK
      DO 240 J=1,NORBL
      TS(I,J) = 0.0
      DO 250 KK=1,NORBK
      TS(I,J) = TS(I,J) + T(I,KK)*TEMP(KK,J)
  250 CONTINUE
  240 CONTINUE
C     FILL BETA MATRIX
      DO 260 I=1,NORBK
      LLKP = KI + I - 1
      DO 260 J=1,NORBL
      LLLP = KJ + J - 1
      BETAO(LLKP,LLLP) = TS(I,J)*BETA(II,JJ)
  260 CONTINUE
C     SYMMETRIZATION OF BETAO MATRIX
      DO 330 I=1,N
      DO 330 J=1,N
  330 BETAO(J,I) = BETAO(I,J)
      KJ = KJ + NORBL
      RETURN
      END
```

5 PROGRAM LISTINGS

```
      SUBROUTINE BET2
C     BEGINNING OF THE SCF ITERATION
C     PO IS THE ORBITAL BOND ORDER - CHARGE DENSITY MATRIX
C     P IS THE ATOMIC BOND ORDER - CHARGE DENSITY MATRIX
      COMMON /ONE/ N,NA,NOCC,ICHGE,IX,IY,I,J,KI,KJ,ISUB,IDUMB,ISTOP,
     1E,TCONV,MU,MO
      COMMON/TWO/ AII(100),AIII(100),XC(35),YC(35),ZC(35),A(100,100),
     1            BETAO(100,100),C(100,100)
      COMMON/THREE/ NAT(35),S(15),ZN(35),ZCHG(35),Q(35)
      COMMON/FOUR/ P(35,35),PI(35,35)
      COMMON/FIVE/ PO(100,100),POI(100,100)
      COMMON/SIX/ U(35,3),R(35,35),BETA(35,35),GAMMA(35,35)
      COMMON/GEOD/ EE(3),T(9,9),TS(9,9),SA(20)
      DIMENSION TC(100)
  102 FORMAT(/5X,28H CONVERGENCE ATTAINED AFTER , I3, 11H ITERATIONS/)
  106 FORMAT(34H JOB ABANDONED AFTER 50 ITERATIONS)
      DO 99 I = 1,NA
      DO 99 J = 1,NA
      P(I,J) = 0.0
   99 PI(I,J) = 0.0
      NI = 0
  299 NI = NI + 1
      IF(200 - NI) 300, 301, 301
  300 WRITE(IY,106)
      TCONV = 0.0
      GO TO 325
  301 DO 302 I = 1,N
  302 TC(I) = AII(I)
      IF (NI-1) 303,303,315
  303 DO 304 I=1,N
      DO 304 J=1,N
  304 PO(I,J) =0.0
      K=1
      DO 314 I=1,NA
      IF (NAT(I)-1) 305,305,306
  305 PO(K,K)=1.0
      P(I,I)=1.0
      K = K + 1
      GO TO 314
  306 IF(NAT(I).GT.10) GO TO 503
      KD = NAT(I)-4
      GO TO (900,307,309,311,400),KD
  900 DO 901 JA=1,4
      PO(K,K) = 0.75
  901 K = K+1
      P(I,I) = 3.0
      GO TO 314
  307 DO 308 JA = 1,4
      PO(K,K) = 1.0
  308 K = K + 1
      P(I,I) = 4.0
      GO TO 314
  309 PO(K,K) = 2.0
      K = K + 1
      DO 310 JA = 1,3
      PO(K,K) = 1.0
  310 K = K + 1
      P(I,I) = 5.0
      GO TO 314
```

```
311 DO 312 JA = 1,2
    PO(K,K) = 2.0
312 K = K + 1
    DO 313 JA = 1,2
    PO(K,K) = 1.0
313 K = K + 1
    P(I,I) = 6.0
    GO TO 314
400 DO 401 JA=1,3
    PO(K,K) = 2.0
401 K = K+1
    PO(K,K) =1.0
    P(I,I) =7.0
    K = K+1
    GO TO 314
503 KD = NAT(I)-12
    GO TO (800,801,802,803,804),KD
800 DO 805 JA=1,4
    PO(K,K) = 0.75
805 K = K+1
    P(I,I) = 3.0
    GO TO 314
801 DO 806 JA=1,4
    PO(K,K) = 1.0
806 K = K+1
    P(I,I) = 4.0
    DO 87 JA=1,5
    PO(K,K) = 0.0
 87 K = K+1
    GO TO 314
802 PO(K,K) = 2.0
    K = K+1
    DO 807 JA=1,3
    PO(K,K) = 1.0
807 K = K+1
    P(I,I) = 5.0
    DO 899 JA=1,5
    PO(K,K) = 0.0
899 K = K+1
    GO TO 314
803 DO 808 JA=1,2
    PO(K,K) = 2.0
808 K = K+1
    DO 809 JA=1,2
    PO(K,K) = 1.0
809 K = K+1
    P(I,I) = 6.0
    DO 896 JA=1,5
    PO(K,K) =0.0
896 K = K+1
    GO TO 314
804 DO 810 JA=1,3
    PO(K,K) = 2.0
810 K = K+1
    PO(K,K) = 1.0
    P(I,I) = 7.0
    K = K+1
    DO 895 JA=1,5
    PO(K,K) = 0.0
```

```
    895 K = K+1
    314 CONTINUE
        DO 500 I=1,N
        DO 500 J=1,N
    500 POI(I,J) = PO(I,J)
        DO 501 I=1,NA
        DO 501 J=1,NA
    501 PI(I,J) = P(I,J)
        GO TO 322
    315 K = 1
        DO 321 I = 1,NA
        IF(NAT(I) - 1) 316,316,318
    316 DO 317 J = 1,N
        SUM = 0.0
        DO 337 L = 1,NOCC
    337 SUM = SUM + 2. * C(L,K) * C(L,J)
    317 PO(K,J) = SUM
        P(I,I) = PO(K,K)
        K = K + 1
        GO TO 321
    318 IF(NAT(I)-10) 850,850,851
    850 SUMT = 0.0
        DO 320 KP = 1,4
        DO 339 J = 1,N
        SUM = 0.0
        DO 319 L = 1,NOCC
    319 SUM = SUM + 2. * C(L,K) * C(L,J)
    339 PO(K,J) = SUM
        SUMT = SUMT + PO(K,K)
    320 K = K + 1
        P(I,I) = SUMT
        GO TO 321
    851 SUMT = 0.0
        DO 860 KP=1,9
        DO 861 J=1,N
        SUM = 0.0
        DO 862 L=1,NOCC
    862 SUM = SUM + 2.0*C(L,K)*C(L,J)
    861 PO(K,J) = SUM
        SUMT = SUMT + PO(K,K)
    860 K = K+1
        P(I,I) = SUMT
    321 CONTINUE
        IF (ISTOP.GT.0) CALL CONVRG
    322 CALL EMOFF
        CALL EMON
        CALL MATRIX (A,C,N,100,0)
        DO 40 I=1,N
 40     AII(I)=A(I,I)
        DO 323 I = 1,N
        IF(ABS(TC(I) - AII(I)) .GT. TCONV) GO TO 299
    323 CONTINUE
        WRITE(IY,102) NI
        IF(ISUB) 325,325,333
    333 WRITE(IY,103)
    103 FORMAT (//33H MOLECULAR ORBITALS AND ENERGIES.)
        CALL PROUT (A,C,N,100)
    325 RETURN
        END
```

```
      SUBROUTINE CONVRG
C     SUBROUTINE CONVERGE
      COMMON /ONE/ N,NA,NOCC,ICHGE,IX,IY,I,J,KI,KJ,ISUB,IDUMB,ISTOP,
     1E,TCONV,MU,MO
      COMMON/FOUR/ P(35,35),PI(35,35)
      COMMON/FIVE/ PO(100,100),POI(100,100)
      AA = 0.5
      BB = 1.0 - AA
      DO 1 I=1,N
      DO 1 J=1,N
      AB = PO(I,J)
      PO(I,J) = BB*PO(I,J) + AA*POI(I,J)
    1 POI(I,J) = AB
      DO 2 I=1,NA
      DO 2 J=1,NA
      AB = P(I,J)
      P(I,J) = BB*P(I,J) + AA*PI(I,J)
    2 PI(I,J) = AB
      RETURN
      END

      SUBROUTINE EMOFF
C     CALCULATION OF THE OFF-DIAGONAL ELEMENTS OF THE ENERGY MATRIX
      COMMON /ONE/ N,NA,NOCC,ICHGE,IX,IY,I,J,KI,KJ,ISUB,IDUMB,ISTOP,
     1E,TCONV,MU,MO
      COMMON/TWO/ AII(100),AIII(100),XC(35),YC(35),ZC(35),A(100,100),
     1            BETAO(100,100),C(100,100)
      COMMON/THREE/ NAT(35),S(15),ZN(35),ZCHG(35),Q(35)
      COMMON/FOUR/ P(35,35),PT(35,35)
      COMMON/FIVE/ PO(100,100),POI(100,100)
      COMMON/SIX/ U(35,3),R(35,35),BETA(35,35),GAMMA(35,35)
      COMMON/SEVEN/ KHARGE,IMULT,NFIRST(35),NLAST(35)
      COMMON/INDO/ CINDO(9,9,18)
      DIMENSION G1(18),G3(18),F2(18),F4(18)
      G1(3) =.092012
      G1(4) =.1407
      G1(5) =.199265
      G1(6) =.267708
      G1(7) =.346029
      G1(8) =.43423
      G1(9) =.532305
      G1(13) = 0.123440
      G1(14)=0.176847
      G1(15)=0.126792
      G1(16)=0.113010
      G1(17)=0.105255
      G3(4)=0.0
      G3(5)=0.0
      G3(6)=0.0
      G3(7)=0.0
      G3(8)=0.0
      G3(9)=0.0
      G3(13) = 0.0
      G3(14)=0.103186
      G3(15)=0.075692
      G3(16)=0.067464
      G3(17)=0.062835
      F2(3) =.049865
      F2(4) =.089125
```

```
      F2(5)  =.13041
      F2(6)  =.17372
      F2(7)  =.219055
      F2(8)  =.266415
      F2(9)  =.31580
      F2(13) = 0.058873
      F2(14)=0.083131
      F2(15)=0.108306
      F2(16)=0.166740
      F2(17)=0.193936
      F4(3)=0.0
      F4(4)=0.0
      F4(5)=0.0
      F4(6)=0.0
      F4(7)=0.0
      F4(8)=0.0
      F4(9)=0.0
      F4(13) = 0.0
      F4(14)=0.054216
      F4(15)=0.070634
      F4(16)=0.108743
      F4(17)=0.126480
      KI = 1
      KJ = 1
      DO 717 I = 1,NA
      DO 714 J = 1,NA
      IF(I-J) 705, 701, 705
  701 IF(NAT(J)-1) 703,703,774
  703 KJ = KJ + 1
      GO TO 714
  774 IF(NAT(J)-10) 704,704,903
  704 KJ = KJ + 4
      GO TO 714
  903 KJ = KJ+9
      GO TO 714
  705 KD = NAT(I) + NAT(J)
      IF(KD - 2) 707, 706, 707
  706 A(KI,KJ) = BETAO(KI,KJ) -.5 *PO(KI,KJ) * GAMMA(I,J)
      KJ = KJ + 1
      GO TO 714
  707 IF(NAT(J)-1) 710,768,710
  768 IF(NAT(I)-10) 708,708,904
  904 DO 905 KIND=1,9
      KT = KI + KIND -1
  905 A(KT,KJ) = BETAO(KT,KJ) - 0.5*PO(KT,KJ)*GAMMA(I,J)
      KJ = KJ+1
      GO TO 714
  708 DO 709 KIND = 1,4
      KT = KI + KIND - 1
  709 A(KT,KJ) = BETAO(KT,KJ) -.5 *PO(KT,KJ) * GAMMA(I,J)
      KJ = KJ + 1
      GO TO 714
  710 IF(NAT(I) - 1) 713, 711, 713
  711 IF(NAT(J).GT.10) GO TO 906
      DO 712 KIND = 1,4
      KT = KJ + KIND - 1
  712 A(KI,KT)= BETAO(KI,KT) -.5* PO(KI,KT) * GAMMA(I,J)
      KJ = KJ + 4
      GO TO 714
```

```
906 DO 907 KIND=1,9
    KT = KJ + KIND -1
907 A(KI,KT) = BETAO(KI,KT) - 0.5*PO(KI,KT)*GAMMA(I,J)
    KJ = KJ+9
    GO TO 714
713 IF(NAT(I).GT.10.AND.NAT(J).GT.10) GO TO 908
    IF(NAT(I).GT.10) GO TO 909
    IF(NAT(J).GT.10) GO TO 910
    DO 720 KIND=1,4
    KT = KI + KIND - 1
    DO 720 LIND = 1,4
    LT = KJ + LIND - 1
720 A(KT,LT) = BETAO(KT,LT) - .5 *PO(KT,LT) * GAMMA(I,J)
    KJ = KJ + 4
    GO TO 714
908 DO 911 KIND=1,9
    KT = KI + KIND -1
    DO 911 LIND=1,9
    LT = KJ + LIND -1
911 A(KT,LT) = BETAO(KT,LT) - 0.5*PO(KT,LT)*GAMMA(I,J)
    KJ = KJ +9
    GO TO 714
909 DO 912 KIND=1,9
    KT = KI + KIND -1
    DO 912 LIND=1,4
    LT = KJ + LIND -1
912 A(KT,LT) = BETAO(KT,LT) - 0.5*PO(KT,LT)*GAMMA(I,J)
    KJ = KJ + 4
    GO TO 714
910 DO 913 KIND=1,4
    KT = KI + KIND -1
    DO 913 LIND=1,9
    LT = KJ + LIND -1
913 A(KT,LT) = BETAO(KT,LT) - 0.5*PO(KT,LT)*GAMMA(I,J)
    KJ = KJ+9
714 CONTINUE
    IF(NAT(I) - 1) 716,715, 716
715 KI = KI + 1
    KJ = 1
    GO TO 717
716 IF(NAT(I).LT.10) GO TO 914
    KI = KI + 9
    KJ = 1
    GO TO 717
914 KI = KI + 4
    KJ = 1
717 CONTINUE
    KI = 1
    DO 725 I = 1,NA
    IF(NAT(I) - 1) 718, 723, 718
718 IF(NAT(I).LT.10) GO TO 915
    DO 916 JK=1,9
    KT = KI + JK - 1
    DO 916 IK=1,9
    LT = KI + IK - 1
916 A(KT,LT) = -0.5*PO(KT,LT)*GAMMA(I,I)
    KI = KI+9
    GO TO 725
915 DO 721 JK=1,4
```

```
  314 CONTINUE
      KT = KI + JK -1
      DO 721 IK=1,4
      LT = KI + IK -1
  721 A(KT,LT) = -.5 * PO(KT,LT) * GAMMA(I,I)
      KI = KI + 4
      GO TO 725
  723 KI = KI + 1
  725 CONTINUE
C     INDO MODIFICATION
      DO 80 II= 1,NA
      K = NAT(II)
      I = NFIRST(II)
      IF(K.EQ.1) GO TO 80
      IM = I-1
      NORB = NLAST(II)-NFIRST(II)+1
      DO 75 JJ=1,NORB
      IJ = IM+JJ
      DO 75 KK=JJ,NORB
      IK = IM+KK
      IF(IJ.EQ.IK) GO TO 75
      A(IK,IJ) = A(IK,IJ) + PO(IJ,IK)*(1.5*CINDO(KK,JJ,K)
     1                              - 0.5*CINDO(JJ,KK,K))
   75 CONTINUE
   80 CONTINUE
      RETURN
      END

      SUBROUTINE EMON
C     CALCULATION OF THE DIAGONAL ELEMENTS OF THE ENERGY MATRIX
      COMMON /ONE/ N,NA,NOCC,ICHGE,IX,IY,I,J,KI,KJ,ISUB,IDUMB,ISTOP,
     1E,TCONV,MU,MO
      COMMON/TWO/ AII(100),AIII(100),XC(35),YC(35),ZC(35),A(100,100),
     1            BETAO(100,100),C(100,100)
      COMMON/THREE/ NAT(35),S(15),ZN(35),ZCHG(35),Q(35)
      COMMON/FOUR/ P(35,35),PI(35,35)
      COMMON/FIVE/ PO(100,100),POI(100,100)
      COMMON/SIX/ U(35,3),R(35,35),BETA(35,35),GAMMA(35,35)
      COMMON/SEVEN/ KHARGE,IMULT,NFIRST(35),NLAST(35)
      COMMON/INDO/ CINDO(9,9,18)
      DIMENSION G1(18),G3(18),F2(18),F4(18)
      G1(3) =.092012
      G1(4) =.1407
      G1(5) =.199265
      G1(6) =.267708
      G1(7) =.346029
      G1(8) =.43423
      G1(9) =.532305
      G1(13) = 0.123440
      G1(14)=0.176847
      G1(15)=0.126792
      G1(16)=0.113010
      G1(17)=0.105255
      G3(4)=0.0
      G3(5)=0.0
      G3(6)=0.0
      G3(7)=0.0
      G3(8)=0.0
      G3(9)=0.0
```

```
      G3(13) = 0.0
      G3(14)=0.103186
      G3(15)=0.075692
      G3(16)=0.067464
      G3(17)=0.062835
      F2(3) =.049865
      F2(4) =.089125
      F2(5) =.13041
      F2(6) =.17372
      F2(7) =.219055
      F2(8) =.266415
      F2(9) =.31580
      F2(13) = 0.058873
      F2(14)=0.083131
      F2(15)=0.108306
      F2(16)=0.166740
      F2(17)=0.193936
      F4(3)=0.0
      F4(4)=0.0
      F4(5)=0.0
      F4(6)=0.0
      F4(7)=0.0
      F4(8)=0.0
      F4(9)=0.0
      F4(13) = 0.0
      F4(14)=0.054216
      F4(15)=0.070634
      F4(16)=0.108743
      F4(17)=0.126480
      KI = 1
      DO 750 I = 1,NA
      SUMP = 0.0
      DO 701 JK = 1,NA
      IF(I.EQ.JK)GO TO 701
      SUMP = SUMP + (P(JK,JK) - ZCHG(JK)) *GAMMA(I,JK)
  701 CONTINUE
      IF(NAT(I) - 1) 740, 720, 740
  720 A(KI,KI) =-.5 * U(I,1) +(P(I,I) -ZCHG(I) -.5 * (PO(KI,KI) -1.))
     1*GAMMA(I,I) + SUMP
      KI = KI + 1
      GO TO 750
  740 A(KI,KI) =-.5 * U(I,1) +(P(I,I) -ZCHG(I) -.5 * (PO(KI,KI) -1.))*
     1GAMMA(I,I) + SUMP
      DO 745 LK = 1,3
      IT = KI + LK
      A(IT,IT) =-.5 *U(I,2) + (P(I,I) -ZCHG(I) -.5 *(PO(IT,IT) -
     11.)) * GAMMA(I,I) + SUMP
  745 CONTINUE
      IF(NAT(I).LT.10) GO TO 751
      DO 752 MK=4,8
      ID = KI + MK
      A(ID,ID) =-0.5*U(I,3) + (P(I,I)-ZCHG(I)-0.5*(PO(ID,ID)-1))
     1                       *GAMMA(I,I) + SUMP
  752 CONTINUE
      KI = KI + 9
      GO TO 750
  751 KI = KI + 4
  750 CONTINUE
C     INDO MODIFICATION
```

```
      DO 80 II=1,NA
      K = NAT(II)
      I = NFIRST(II)
      IF(K.EQ.1) GO TO 80
      IM = I-1
      NORB = NLAST(II) - NFIRST(II) + 1
      DO 75 JJ=1,NORB
      IJ = IM+JJ
      ZZ = 0.0
      DO 75 KK=JJ,NORB
      IK = IM+KK
      IF(IJ.NE.IK) GO TO 75
      DO 65 LL=1,NORB
      IF(JJ.GT.LL) GO TO 55
      MM = JJ
      NN = LL
      GO TO 60
   55 MM = LL
      NN = JJ
   60 CONTINUE
      IL = IM+LL
   65 ZZ = ZZ + PO(IL,IL)*(CINDO(MM,NN,K) - 0.5*CINDO(NN,MM,K))
      A(IJ,IJ) = A(IJ,IJ) + ZZ
   75 CONTINUE
   80 CONTINUE
      DO 280 I=1,NA
      K = NAT(I)
      J = NFIRST(I)
      NORB = NLAST(I) - NFIRST(I) + 1
      IF(K.LT.3) GO TO 280
      IF(K.EQ.3.OR.K.EQ.11) GO TO 220
      A(J,J) = A(J,J) + (ZCHG(I)-1.5)*G1(K)/6.0
      IF(K.EQ.4.OR.K.EQ.12) GO TO 240
      TEMP = G1(K)/3.0 + (ZCHG(I)-2.5)*2.0*F2(K)/25.0
      TEMD = F2(K)/5.0 + (ZCHG(I)-2.5)*(G1(K)/15.0 + 3.0*G3(K)/70.0)
      GO TO 260
  220 TEMP = G1(K)/12.0
      TEMD = F2(K)/20.0
      GO TO 260
  240 TEMP = G1(K)/4.0
      TEMD = 3.0*F2(K)/20.0
  260 CONTINUE
      DO 270 L=1,3
      JPL = J+L
  270 A(JPL,JPL) = A(JPL,JPL) + TEMP
      IF(NORB.NE.9) GO TO 280
      DO 275 JJ=1,5
      JJJ = JPL+JJ
  275 A(JJJ,JJJ) = A(JJJ,JJJ) + TEMD
  280 CONTINUE
      RETURN
      END
```

THEORY OF NMR PARAMETERS

```
      SUBROUTINE INDOMO
      COMMON/ONE/ N,NA,NOCC,ICHGE,IX,IY,II,JJ,KI,KJ,ISUB,IDUMB,ISTOP,
     1            E,TCONV,MU,MO
      COMMON/TWO/ AII(100),AIII(100),XC(35),YC(35),ZC(35),CIM(10000),
     1            AJT(100),AJJ(100),ATT(100),XX(100),DUMMY(9600),
     2            C(100,100)
      COMMON/THREE/ NAT(35),S(15),ZN(35),ZCHG(35),Q(35)
      COMMON/SEVEN/ KHARE,IMULT,NFIRST(35),NLAST(35)
      COMMON/INDO/ CINDO(9,9,18)
      DIMENSION G1(18),G3(18),F2(18),F4(18),KN(18)
      G1(3)=0.092012
      G1(4)=0.140700
      G1(5)=0.199265
      G1(6)=0.267708
      G1(7)=0.346029
      G1(8)=0.434230
      G1(9)=0.532305
      G1(13) = 0.123440
      G1(14)=0.176847
      G1(15)=0.126792
      G1(16)=0.113010
      G1(17)=0.105255
      G3(4)=0.0
      G3(5)=0.0
      G3(6)=0.0
      G3(7)=0.0
      G3(8)=0.0
      G3(9)=0.0
      G3(13) = 0.0
      G3(14)=0.103186
      G3(15)=0.075692
      G3(16)=0.067464
      G3(17)=0.062835
      F2(3)=0.049865
      F2(4)=0.089125
      F2(5)=0.130410
      F2(6)=0.173720
      F2(7)=0.219055
      F2(8)=0.266415
      F2(9)=0.315800
      F2(13) = 0.058873
      F2(14)=0.083131
      F2(15)=0.108306
      F2(16)=0.166740
      F2(17)=0.193936
      F4(3)=0.0
      F4(4)=0.0
      F4(5)=0.0
      F4(6)=0.0
      F4(7)=0.0
      F4(8)=0.0
      F4(9)=0.0
      F4(13) = 0.0
      F4(14)=0.054216
      F4(15)=0.070634
      F4(16)=0.108743
      F4(17)=0.126480
      DO 300 K=1,18
  300 KN(K) =0
```

```
      DO 1000 M=1,NA
      K = NAT(M)
      KOK = KN(K)
      IF(KOK.NE.0) GO TO 1000
      KN(K) = KN(K)+1
      DO 320 J=1,9
      DO 310 I=1,J
      CINDO(I,J,K) = 0.0
  310 CINDO(J,I,K) = CINDO(I,J,K)
  320 CONTINUE
      IF(K.EQ.1) GO TO 1000
      DO 900 J=1,9
      DO 800 I=1,4
      IF(I.NE.J) GO TO 330
      IF(I.EQ.1) GO TO 800
      IF(I.LT.5) GO TO 350
C     THE DIAGONAL DD/DD ELEMENTS
      CINDO(I,I,K) = 4.0*F2(K)/49.0 + 36.0*F4(K)/441.0
      GO TO 800
C     THE DIAGONAL PP/PP ELEMENTS
  350 CINDO(I,I,K) = 4.0*F2(K)/25.0
      GO TO 800
  330 CONTINUE
      IF(I.NE.1) GO TO 710
      IF(I.LT.5) GO TO 700
C     NON-DIAGONAL SD/SD ELEMENTS AND SS/DD IS ZERO
      CINDO(J,I,K) = F2(K)/5.0
      GO TO 800
C     NON-DIAGONAL SP/SP ELEMENTS AND SS/PP IS ZERO
  700 CINDO(J,I,K) = G1(K)/3.0
      GO TO 800
  710 CONTINUE
      IF(I.GT.4) GO TO 500
      IF(J.GT.4) GO TO 600
C     NON-DIAGONAL PP/PP ELEMENTS
      CINDO(I,J,K) = -2.0*F2(K)/25.0
      CINDO(J,I,K) = 3.0*F2(K)/25.0
      GO TO 800
  600 CONTINUE
      IF(J.NE.5) GO TO 400
      IF(I.EQ.4) GO TO 620
C     P(X)/D(Z**2) AND P(Y)/D(Z**2) ELEMENTS
      CINDO(I,J,K) = -2.0*F2(K)/35.0
      CINDO(J,I,K) = G1(K)/15.0 + 18.0*G3(K)/245.0
      GO TO 800
C     P(Z)/D(Z**2) ELEMENTS
  620 CINDO(I,J,K) = 4.0*F2(K)/35.0
      CINDO(J,I,K) = 4.0*G1(K)/15.0 + 27.0*G3(K)/245.0
      GO TO 800
  400 CONTINUE
C     THE REST OF P/D ELEMENTS
      IF(J.GT.7) GO TO 450
      IF(I-3) 410,440,415
  410 IF(J.EQ.7) GO TO 420
  415 CINDO(I,J,K) = 2.0*F2(K)/35.0
      CINDO(J,I,K) = 3.0*G1(K)/15.0 + 24.0*G3(K)/245.0
      GO TO 800
  420 CINDO(I,J,K) = -4.0*F2(K)/35.0
      CINDO(J,I,K) =  3.0*G1(K)/15.0 -  6.0*G3(K)/245.0
```

```
            GO TO 800
    440 IF(J.EQ.6) GO TO 420
            GO TO 415
    450 IF(I.EQ.4) GO TO 460
            CINDO(I,J,K) = 2.0*F2(K)/35.0
            CINDO(J,I,K) = 6.0*G1(K)/15.0 - 3.0*G3(K)/245.0
            GO TO 800
    460 CINDO(I,J,K) = -4.0*F2(K)/35.0
            CINDO(J,I,K) = 15.0*G3(K)/245.0
            GO TO 800
    500 CONTINUE
C       NON-DIAGONAL DD/DD ELEMENTS
            IF(I.NE.5) GO TO 520
            IF(J.GT.7) GO TO 510
            CINDO(I,J,K) = 2.0*F2(K)/49.0 - 24.0*F4(K)/441.0
            CINDO(J,I,K) =     F2(K)/49.0 + 30.0*F4(K)/441.0
    510 CINDO(I,J,K) = -4.0*F2(K)/49.0 + 6.0*F4(K)/441.0
            CINDO(J,I,K) =  4.0*F2(K)/49.0 + 15.0*F4(K)/441.0
            GO TO 800
    520 IF(I.EQ.8) GO TO 530
            CINDO(I,J,K) = -2.0*F2(K)/49.0 - 4.0*F4(K)/441.0
            CINDO(J,I,K) =  3.0*F2(K)/49.0 + 20.0*F4(K)/441.0
            GO TO 800
    530 CINDO(I,J,K) = 4.0*F2(K)/49.0  34.0*F4(K)/441.0
            CINDO(J,I,K) = 35.0*F4(K)/441.0
    800 CONTINUE
    900 CONTINUE
   1000 CONTINUE
            RETURN
            END

            SUBROUTINE REPS
            COMMON/ONE/ N,NA,NOCC,ICHGE,IX,IY,I,J,KI,KJ,ISUB,IDUMB,ISTOP,
           1E,TCONV,MU,MO
            COMMON/THREE/ NAT(35),S(15),ZN(35),ZCHG(35),Q(35)
            COMMON/FOUR/ P(35,35),PI(35,35)
            COMMON/FIVE/ PO(100,100),REP(100,100)
            COMMON/SIX/ U(35,3),R(35,35),BETA(35,35),GAMMA(35,35)
            II=1
            DO 13 I=1,NA
            JJ=1
            DO 10 J=1,NA
            K=NAT(I)+NAT(J)
            IF (K-2)1,1,2
      1 REP(II,JJ)=GAMMA(I,J)
            JJ=JJ+1
            GO TO 10
      2     IF (NAT(I)-1)3,3,5
         3  IF(NAT(J).LT.10) GO TO 20
            DO 21 JK=1,9
            JJJ = JJ + JK -1
     21 REP(II,JJJ) = GAMMA(I,J)
            JJ = JJ + 9
            GO TO 10
     20 DO 4 JK=1,4
            JJJ=JJ+JK-1
      4 REP(II,JJJ)=GAMMA(I,J)
            JJ=JJ+4
            GO TO 10
```

```
    5     IF(NAT(J)-1)6,6,8
    6     IF(NAT(I).LT.10) GO TO 22
          DO 23 IK=1,9
          III = II + IK -1
   23     REP(III,JJ) = GAMMA(I,J)
          JJ = JJ +1
          GO TO 10
   22     DO 7 IK=1,4
          III=II+IK-1
    7     REP(III,JJ)=GAMMA(I,J)
          JJ=JJ+1
          GO TO 10
    8     IF((NAT(I).GT.10).OR.(NAT(J).GT.10)) GO TO 24
          DO 9 JK=1,4
          JJJ=JJ+JK-1
          DO 9 IK=1,4
          III=II+IK-1
    9     REP(III,JJJ)=GAMMA(I,J)
          JJ=JJ+4
          GO TO 10
   24     IF((NAT(I).GT.10).AND.(NAT(J).GT.10)) GO TO 25
          IF(NAT(J).GT.10) GO TO 26
          DO 27 IK=1,9
          III = II + IK -1
   27     REP(III,JJ) = GAMMA(I,J)
          JJ = JJ +4
          GO TO 10
   26     DO 28 JK=1,9
          JJJ = JJ + JK -1
   28     REP(II,JJJ) = GAMMA(I,J)
          JJ = JJ +9
          GO TO 10
   25     DO 29 JK=1,9
          JJJ = JJ + JK -1
          DO 29 IK=1,9
          III = II + IK -1
   29     REP(III,JJJ) = GAMMA(I,J)
          JJ = JJ + 9
   10     CONTINUE
          IF (NAT(I)-1) 11,11,12
   11     II=II+1
          GO TO 13
   12     IF(NAT(I).GT.10) GO TO 30
          II = II +4
          GO TO 13
   30     II = II + 9
   13     CONTINUE
          RETURN
          END
```

```
      SUBROUTINE TRANS(I,J,A1,A2,A3,B1,B2,B3)
      COMMON/TYPE/ ITYPE(100),XTOY(5,8),XTYPE(18)
      IF(I.EQ.0) GO TO 100
      XTOY(I,1) = XTOY(I,1) + A1
      XTOY(I,2) = XTOY(I,2) + A2
      XTOY(I,3) = XTOY(I,3) + A3
      XTOY(I,4) = XTOY(I,4) + (A1+A2+A3)/3.0
      IF(J.EQ.0) GO TO 100
      XTOY(I,5) = XTOY(I,5) + B1
      XTOY(I,6) = XTOY(I,6) + B2
      XTOY(I,7) = XTOY(I,7) + B3
      XTOY(I,8) = XTOY(I,8) + (B1+B2+B3)/3.0
  100 RETURN
      END

      SUBROUTINE CIMAT
      COMMON/ONE/ N,NA,NOCC,ICHGE,IX,IY,I,J,KI,KJ,ISUB,IDUMB,ISTOP,
     1E,TCONV,MU,MO
      COMMON/TWO/ AII(100),AIII(100),XC(35),YC(35),ZC(35),CIM(10000),
     1            AJT(100),AJJ(100),ATT(100),XX(100),DUMMY(9600),
     2            C(100,100)
      COMMON/THREE/ NAT(35),S(15),ZN(35),ZCHG(35),Q(35)
      COMMON/FOUR/ P(35,35),PI(35,35)
      COMMON /FIVE/ PO(100,100),REP(100,100)
      COMMON/SIX/ U(35,3),R(35,35),BETA(35,35),GAMMA(35,35)
      MO = NOCC
      MU = N-NOCC
   22 MOO=NOCC-MO+1
      MUU=NOCC+MU
      KK=MU*MO
      M=N-1
      DO 8 I=1,KK
      B=AMOD(FLOAT(I)/FLOAT(MU),1.0)
      IF(B.LT.0.001) GO TO 1
      MJ=I/MU+NOCC-MO+1
      MT=(B+0.01)*MU+NOCC
      GO TO 2
    1 MJ=I/MU +NOCC-MO
      MT=MU+NOCC
    2 DO 3 NQ=1,N
      AJT(NQ) = C(MJ,NQ)*C(MT,NQ)
      AJJ(NQ) = C(MJ,NQ)*C(MJ,NQ)
      ATT(NQ) = C(MT,NQ)*C(MT,NQ)
    3 CONTINUE
      SUM1 = 0.0
      SUM3 = 0.0
      DO 7 K=1,N
      DO 7 L=1,N
      SUM1 = SUM1 + AJT(K)*REP(K,L)*AJT(L)
      SUM3 = SUM3 + AJJ(K)*REP(K,L)*ATT(L)
    7 CONTINUE
C     INDO MODIFICATION
      SUM1 = SUM1 + CORR(MJ,MT,MJ,MT)
      SUM3 = SUM3 + CORR(MJ,MJ,MT,MT)
      B = AII(MT)-AII(MJ)
      CIM(I) = ( B + 2.0*SUM1 - SUM3)*27.204
    8 CONTINUE
      RETURN
      END
```

```
      FUNCTION CORR(I,J,K,L)
C     CALCULATE CORRECTIONS TO THE CNDO MOLECULAR INTEGRALS
C     TO CONVERT THE CI MATRIX ELEMENTS ACCORDING TO INDO -
      COMMON/ONE/ N,NA,NOCC,ICHGE,IX,IY,II,JJ,KI,KJ,ISUB,IDUMB,ISTOP,
     1E,TCONV,MU,MO
      COMMON/TWO/ AII(100),AIII(100),XC(35),YC(35),ZC(35),CIM(10000),
     1            AJT(100),AJJ(100),ATT(100),XX(100),DUMMY(9600),
     2            C(100,100)
      COMMON/THREE/ NAT(35),S(15),ZN(35),ZCHG(35),Q(35)
      COMMON/SEVEN/ KHARGE,IMULT,NFIRST(35),NLAST(35)
      DIMENSION G1(18),G3(18),F2(18),F4(18)
      G1(3) =0.092012
      G1(4) =0.1407
      G1(5) =0.199265
      G1(6) =0.267708
      G1(7) =0.346029
      G1(8) =0.43423
      G1(9) =0.532305
      G1(13) = 0.123440
      G1(14)=0.176847
      G1(15)=0.126792
      G1(16)=0.113010
      G1(17)=0.105255
      G3(4)=0.0
      G3(5)=0.0
      G3(6)=0.0
      G3(7)=0.0
      G3(8)=0.0
      G3(9)=0.0
      G3(13) = 0.0
      G3(14)=0.103186
      G3(15)=0.075692
      G3(16)=0.067464
      G3(17)=0.062835
      F2(3) =0.049865
      F2(4) =0.089125
      F2(5) =0.13041
      F2(6) =0.17372
      F2(7) =0.219055
      F2(8) =0.266415
      F2(9) =0.31580
      F2(13) = 0.058873
      F2(14)=0.083131
      F2(15)=0.108306
      F2(16)=0.166740
      F2(17)=0.193936
      F4(3)=0.0
      F4(4)=0.0
      F4(5)=0.0
      F4(6)=0.0
      F4(7)=0.0
      F4(8)=0.0
      F4(9)=0.0
      F4(13) = 0.0
      F4(14)=0.054216
      F4(15)=0.070634
      F4(16)=0.108743
      F4(17)=0.126480
      CORCTN =0.0
```

```
      DO 70  IA=1,NA
      NI = NAT(IA)
      IF(NI.EQ.1) GO TO 70
      LS = NFIRST(IA)
      LX = LS+1
      LY = LS+2
      LZ = LS+3
      IF(NI.LT.10) GO TO 99
      LDA = LS+4
      LDB = LS+5
      LDC = LS+6
      LDD = LS+7
      LDE = LS+8
 99   CONTINUE
      DO 10  M=LX,LZ
      CORCTN = CORCTN +(C(I,M)*C(J,LS) + C(I,LS)*C(J,M))
     1 *(C(K,M)*C(L,LS) + C(K,LS)*C(L,M)) * G1(NI)/3.0
 10   CONTINUE
      DO 30 M=LX,LY
      M1 = M +1
      DO  20   LM=M1,LZ
      CORCTN = CORCTN +(C(I,M)*C(J,LM)+C(I,LM)*C(J,M))
     1 *(C(K,M)*C(L,LM) + C(K,LM)*C(L,M)) * 3.0 * F2(NI)/25.
 20   CONTINUE
 30   CONTINUE
      DO 40  M= LX,LZ
      CORCTN =CORCTN +(C(I,M)*C(J,M)*C(K,M)*C(L,M))*4.0 * F2(NI)/25.0
 40   CONTINUE
      DO 60  M = LX,LY
      M1 =M + 1
      DO 50   LM=M1,LZ
      CORCTN = CORCTN - (C(I,M)*C(J,M)*C(K,LM)*C(L,LM) + C(I,LM)*
     1 C(J,LM)*C(K,M)*C(L,M)) * 2.0 * F2(NI)/25.0
 50   CONTINUE
 60   CONTINUE
      IF(NI.LT.10) GO TO 70
      DO 110 M=LDA,LDE
      CORCTN = CORCTN+(C(I,M)*C(J,LS)+C(I,LS)*C(J,M))
     1              *(C(K,M)*C(L,LS)+C(K,LS)*C(L,M))*F2(NI)/5.0
 110  CONTINUE
      DO 120 M=LX,LY
      CORCTN = CORCTN+(C(I,M)*C(J,LDA)+C(I,LDA)*C(J,M))
     1              *(C(K,M)*C(L,LDA)+C(K,LDA)*C(L,M))
     2              *(G1(NI)/15.0 + 18.0*G3(NI)/245.0)
 120  CONTINUE
      DO 130 M=LDB,LDB
      CORCTN = CORCTN+(C(I,M)*C(J,LX)+C(I,LX)*C(J,M))
     1              *(C(K,M)*C(L,LX)+C(K,LX)*C(L,M))
     2              *(G1(NI)/5.0 + 24.0*G3(NI)/245.0)
 130  CONTINUE
      DO 131 M=LDD,LDE
      CORCTN = CORCTN+(C(I,M)*C(J,LX)+C(I,LX)*C(J,M))
     1              *(C(K,M)*C(L,LX)+C(K,LX)*C(L,M))
     2              *(G1(NI)/5.0 + 24.0*G3(NI)/245.0)
 131  CONTINUE
      DO 132 M=LDC,LDE
      CORCTN = CORCTN+(C(I,M)*C(J,LY)+C(I,LY)*C(J,M))
     1              *(C(K,M)*C(L,LY)+C(K,LY)*C(L,M))
     2              *(G1(NI)/5.0 + 24.0*G3(NI)/245.0)
```

```
  132 CONTINUE
      DO 133 M=LDB,LDC
      CORCTN = CORCTN+(C(I,M)*C(J,LZ)+C(I,LZ)*C(J,M))
     1               *(C(K,M)*C(L,LZ)+C(K,LZ)*C(L,M))
     2               *(G1(NI)/5.0 + 24.0*G3(NI)/245.0)
  133 CONTINUE
      DO 140 M=LDC,LDC
      CORCTN = CORCTN+(C(I,M)*C(J,LX)+C(I,LX)*C(J,M))
     1               *(C(K,M)*C(L,LX)+C(K,LX)*C(L,M))
     2               *(15.0*G3(NI)/245.0)
  140 CONTINUE
      DO 141 M=LDB,LDB
      CORCTN = CORCTN+(C(I,M)*C(J,LY)+C(I,LY)*C(J,M))
     1               *(C(K,M)*C(L,LY)+C(K,LY)*C(L,M))
     2               *(15.0*G3(NI)/245.0)
  141 CONTINUE
      DO 142 M=LDD,LDE
      CORCTN = CORCTN+(C(I,M)*C(J,LZ)+C(I,LZ)*C(J,M))
     1               *(C(K,M)*C(L,LZ)+C(K,LZ)*C(L,M))
     2               *(15.0*G3(NI)/245.0)
  142 CONTINUE
      DO 150 M=LDA,LDA
      CORCTN = CORCTN+(C(I,M)*C(J,LZ)+C(I,LZ)*C(J,M))
     1               *(C(K,M)*C(L,LZ)+C(K,LZ)*C(L,M))
     2               *(4.0*G1(NI)/15.0 + 27.0*G3(NI)/245.0)
  150 CONTINUE
      DO 160 M=LDB,LDC
      CORCTN = CORCTN+(C(I,M)*C(J,LDA)+C(I,LDA)*C(J,M))
     1               *(C(K,M)*C(L,LDA)+C(K,LDA)*C(L,M))
     2               *(F2(NI)/49.0 + 30.0*F4(NI)/441.0)
  160 CONTINUE
      DO 170 M=LDD,LDE
      CORCTN = CORCTN+(C(I,M)*C(J,LDA)+C(I,LDA)*C(J,M))
     1               *(C(K,M)*C(L,LDA)+C(K,LDA)*C(L,M))
     2               *(4.0*F2(NI)/49.0 + 15.0*F4(NI)/441.0)
  170 CONTINUE
      DO 180 M=LDE,LDE
      CORCTN = CORCTN+(C(I,M)*C(J,LDD)+C(I,LDD)*C(J,M))
     1               *(C(K,M)*C(L,LDD)+C(K,LDD)*C(L,M))
     2               *(35.0*F4(NI)/441.0)
  180 CONTINUE
      DO 190 M=LDB,LDC
      M1 = M + 1
      DO 191 LM=M1,LDE
      CORCTN = CORCTN+(C(I,M)*C(J,LM)+C(I,LM)*C(J,M))
     1               *(C(K,M)*C(L,LM)+C(K,LM)*C(L,M))
     2               *(3.0*F2(NI)/49.0 + 20.0*F4(NI)/441.0)
  191 CONTINUE
  190 CONTINUE
      DO 210 M=LDA,LDE
      CORCTN = CORCTN+(C(I,M)*C(J,M)*C(K,M)*C(L,M))
     1               *(4.0*F2(NI)/49.0 + 36.0*F4(NI)/441.0)
  210 CONTINUE
      DO 220 M=LX,LY
      CORCTN = CORCTN-(C(I,M)*C(J,M)*C(K,LDA)*C(L,LDA)
     1                +C(I,LDA)*C(J,LDA)*C(K,M)*C(L,M))
     2               *(2.0*F2(NI)/35.0)
  220 CONTINUE
      DO 230 M=LDA,LDA
```

```
      CORCTN = CORCTN+(C(I,LZ)*C(J,LZ)*C(K,M)*C(L,M)
     1               +C(I,M)*C(J,M)*C(K,LZ)*C(L,LZ))
     2              *(4.0*F2(NI)/35.0)
  230 CONTINUE
      DO 240 M=LDC,LDC
      CORCTN = CORCTN-(C(I,LX)*C(J,LX)*C(K,M)*C(L,M)
     1               +C(I,M)*C(J,M)*C(K,LX)*C(L,LX))
     2              *(4.0*F2(NI)/35.0)
  240 CONTINUE
      DO 241 M=LDB,LDB
      CORCTN = CORCTN-(C(I,LY)*C(J,LY)*C(K,M)*C(L,M)
     1               +C(I,M)*C(J,M)*C(K,LY)*C(L,LY))
     2              *(4.0*F2(NI)/35.0)
  241 CONTINUE
      DO 242 M=LDD,LDE
      CORCTN = CORCTN-(C(I,LZ)*C(J,LZ)*C(K,M)*C(L,M)
     1               +C(I,M)*C(J,M)*C(K,LZ)*C(L,LZ))
     2              *(4.0*F2(NI)/35.0)
  242 CONTINUE
      DO 250 M=LDB,LDB
      CORCTN = CORCTN+(C(I,LX)*C(J,LX)*C(K,M)*C(L,M)
     1               +C(I,M)*C(J,M)*C(K,LX)*C(L,LX))
     2              *(2.0*F2(NI)/35.0)
  250 CONTINUE
      DO 251 M=LDD,LDE
      CORCTN = CORCTN+(C(I,LX)*C(J,LX)*C(K,M)*C(L,M)
     1               +C(I,M)*C(J,M)*C(K,LX)*C(L,LX))
     2              *(2.0*F2(NI)/35.0)
  251 CONTINUE
      DO 252 M=LDC,LDE
      CORCTN = CORCTN+(C(I,LY)*C(J,LY)*C(K,M)*C(L,M)
     1               +C(I,M)*C(J,M)*C(K,LY)*C(L,LY))
     2              *(2.0*F2(NI)/35.0)
  252 CONTINUE
      DO 253 M=LDB,LDC
      CORCTN = CORCTN+(C(I,LZ)*C(J,LZ)*C(K,M)*C(L,M)
     1               +C(I,M)*C(J,M)*C(K,LZ)*C(L,LZ))
     2              *(2.0*F2(NI)/35.0)
  253 CONTINUE
      DO 260 M=LDB,LDC
      CORCTN = CORCTN+(C(I,LDA)*C(J,LDA)*C(K,M)*C(L,M)
     1               +C(I,M)*C(J,M)*C(K,LDA)*C(L,LDA))
     2              *(2.0*F2(NI)/49.0 - 24.0*F4(NI)/441.0)
  260 CONTINUE
      DO 270 M=LDD,LDE
      CORCTN = CORCTN+(C(I,LDA)*C(J,LDA)*C(K,M)*C(L,M)
     1               +C(I,M)*C(J,M)*C(K,LDA)*C(L,LDA))
     2              *(- 4.0*F2(NI)/49.0 + 6.0*F4(NI)/441.0)
  270 CONTINUE
      DO 280 M=LDE,LDE
      CORCTN = CORCTN+(C(I,LDD)*C(J,LDD)*C(K,M)*C(L,M)
     1               +C(I,M)*C(J,M)*C(K,LDD)*C(L,LDD))
     2              *(4.0*F2(NI)/49.0 - 34.0*F4(NI)/441.0)
  280 CONTINUE
      DO 290 M=LDB,LDC
      M1 = M + 1
      DO 291 LM=M1,LDE
      CORCTN = CORCTN+(C(I,M)*C(J,M)*C(K,LM)*C(L,LM)
     1               +C(I,LM)*C(J,LM)*C(K,M)*C(L,M))
```

```
      2                    *(- 2.0*F2(NI)/49.0 - 4.0*F4(NI)/441.0)
  291 CONTINUE
  290 CONTINUE
   70 CONTINUE
      CORR = CORCTN
      RETURN
      END

      SUBROUTINE SHIFT1
      COMMON/ONE/ N,NA,NOCC,ICHGE,IX,IY,I,J,KI,KJ,ISUB,IDUMB,ISTOP,
     1E,TCONV,MU,MO
      COMMON/TWO/ AII(100),AIII(100),XC(35),YC(35),ZC(35),CIM(10000),
     1            AJT(100),AIS(100),AST(100),AIJ(100),DUMMY(9600),
     2            C(100,100)
      COMMON/THREE/ NAT(35),S(15),ZN(35),ZCHG(35),Q(35)
      COMMON/FOUR/ P(35,35),PI(35,35)
      COMMON/FIVE/ PO(100,100),DIAS(35),DIAB(35),PARAS(35),PARAB(35),
     1             DIACC(35),PARCC(35),GGG(35,9),ZSLTR(35),ZS(35),
     2             ZP(35),PARASE(35),PARABE(35),PARCE(35),RS(35),RB(35
      COMMON/SIX/ U(35,3),R(35,35),BETA(35,35),GAMMA(35,35)
      COMMON/TYPE/ ITYPE(100),XTOY(5,8),GT1(3,3),GT2(3,3)
      DIMENSION IATOMS(35)
      DIMENSION ER(9),EI(9),ITS(9)
      DIMENSION RD(35),GAMMB(35,9),GGGG(35,9)
      DIMENSION TRA(5),ATOM(10),DIA(3,5)
      DATA TRA(1)/8HS TO S* /
      DATA TRA(2)/8HS TO PI*/
      DATA TRA(3)/8HPI TO S*/
      DATA TRA(4)/8HN TO S* /
      DATA TRA(5)/8HN TO PI*/
      DATA DIA(1)/0.0/
      DATA DIA(2)/0.0/
      DATA DIA(3)/0.0/
      DATA DIA(4)/251.0/
      DATA DIA(5)/261.0/
      DATA DIA(6)/268.0/
      DATA DIA(7)/315.0/
      DATA DIA(8)/325.0/
      DATA DIA(9)/332.0/
      DATA DIA(10)/384.0/
      DATA DIA(11)/395.0/
      DATA DIA(12)/403.0/
      DATA DIA(13)/453.0/
      DATA DIA(14)/464.0/
      DATA DIA(15)/472.0/
      DATA ATOM(1)/8H   B    /
      DATA ATOM(2)/8H   C    /
      DATA ATOM(3)/8H   N    /
      DATA ATOM(4)/8H   O    /
      DATA ATOM(5)/8H   F    /
      DATA ATOM(6)/8H   AL   /
      DATA ATOM(7)/8H   SI   /
      DATA ATOM(8)/8H   P    /
      DATA ATOM(9)/8H   S    /
      DATA ATOM(10)/8H   CL   /
      READ(IX,110) (IATOMS(I),I=1,35)
  110 FORMAT(35I2)
      KK=MU*MO
      IS=1HS
```

```
      IP=1HP
      IN=1HN
      JATOM=0
      DO 5000 IT1=1,NA
      IF(IATOMS(IT1).EQ.0) GO TO 5000
      JATOM=1
      GO TO 5500
 5000 CONTINUE
 5500 IF(JATOM.EQ.1) READ(5,5550)(ITYPE(IT1),IT1=1,60)
 5550 FORMAT(60A1)
   82 NAM = NOCC + 1
      KA = 1
      DO 7 IA=1,NA
      DO 4000 IT1=1,5
      DO 4000 IT2=1,8
 4000 XTOY(IT1,IT2)=0.0
      JATOM = 0
      DO 222 IM=1,35
      IF(IA.NE.IATOMS(IM)) GO TO 222
      JATOM = 1
  222 CONTINUE
      IF(JATOM.NE.1) GO TO 223
      IBB = NAT(IA) - 4
      WRITE(6,111) ATOM(IBB),IA
      WRITE(6,112)
  111 FORMAT(///51H CONTRIBUTIONS FROM VARIOUS ELECTRONIC TRANSITIONS ,
     1 6HTO THE,A8,39H SCREENING TENSOR ELEMENTS FOR ATOM NO.,I3/)
  112 FORMAT(7H   I - J,4H  KK,4X,6H EIJ   ,4X,6H EST   ,10H      SXX   ,
     1 10H      SYY,10H      SZZ ,10H PARX(EST),10H PARY(EST),
     2 10H PARZ(EST),10H PARX(EIJ),10H PARY(EIJ),10H PARZ(EIJ)/)
  223 Q(IA) = ZCHG(IA) - P(IA,IA)
      IF(NAT(IA) -1) 2,1,2
    1 KA = KA+1
      GO TO 7
    2 LX = KA+1
      LY = KA+2
      LZ = KA+3
      IF(NAT(IA).GT.10) GO TO 50
      KA = KA+4
      GO TO 51
   50 LDA = KA+4
      LDB = KA+5
      LDC = KA+6
      LDD = KA+7
      LDE = KA+8
      KA = KA+9
      GO TO 52
   51 CONTINUE
      LS = LX - 1
      QP = P(IA,IA) - PO(LS,LS)
      ANIA = NAT(IA)
      ZS(IA) = ANIA - 1.60 - 0.4 * PO(LS,LS) - 0.35 * QP
      ZP(IA) = ANIA - 1.65 - 0.5 * PO(LS,LS) - 0.35 * QP
      ZSLTR(IA) = ( 2.0 * ZN(IA) + 0.35 * Q(IA))
      FW = 2.0 * (ANIA - 0.3)
      DIAS(IA) = 17.7503 * (FW + P(IA,IA) * ZSLTR(IA)/4.0)
      DIAB(IA) = 17.7503*(FW + PO(LS,LS)*ZS(IA)/4.0 + QP*ZP(IA)/4.0)
      ZED = 5.91792 / (ZN(IA)*ZN(IA))
      BETA(IA,1) = -(PO(LX,LX)+PO(LY,LY)*2.0+PO(LZ,LZ)*2.0)*ZED
```

```
      BETA(IA,2) = -(PO(LX,LX)*2.0+PO(LY,LY)+PO(LZ,LZ)*2.0)*ZED
      BETA(IA,3) = -(PO(LX,LX)*2.0+PO(LY,LY)*2.0+PO(LZ,LZ))*ZED
      BETA(IA,4) = 0.0
      BETA(IA,5) = 0.0
      BETA(IA,6) = 0.0
      GO TO 53
   52 CONTINUE
      ZS(IA) = 0.0
      ZP(IA) = 0.0
      ANIA = NAT(IA)
      ZSLTR(IA) = 3.0*ZN(IA) + 0.35*Q(IA)
      FW = 4.0*(ANIA-2.225)
      DIAS(IA) = 17.7503*(FW + P(IA,IA)*ZSLTR(IA)/9.0)
      DIAB(IA) = 0.0
      ZED = 11.04678/(ZN(IA)*ZN(IA))
      BETA(IA,1) = -(PO(LX,LX)+PO(LY,LY)*2.0+PO(LZ,LZ)*2.0)*ZED
      BETA(IA,2) = -(PO(LX,LX)*2.0+PO(LY,LY)+PO(LZ,LZ)*2.0)*ZED
      BETA(IA,3) = -(PO(LX,LX)*2.0+PO(LY,LY)*2.0+PO(LZ,LZ))*ZED
      ZEDD = 2.6301864/(ZN(IA)*ZN(IA))
      BETA(IA,4) = -(PO(LDA,LDA)*8.0
     1              +PO(LDB,LDB)*6.0
     2              +PO(LDC,LDC)*9.0
     3              +PO(LDD,LDD)*6.0
     4              +PO(LDE,LDE)*6.0)*ZEDD
      BETA(IA,5) = -(PO(LDA,LDA)*8.0
     1              +PO(LDB,LDB)*9.0
     2              +PO(LDC,LDC)*6.0
     3              +PO(LDD,LDD)*6.0
     4              +PO(LDE,LDE)*6.0)*ZEDD
      BETA(IA,6) = -(PO(LDA,LDA)*5.0
     1              +PO(LDB,LDB)*6.0
     2              +PO(LDC,LDC)*6.0
     3              +PO(LDD,LDD)*9.0
     4              +PO(LDE,LDE)*9.0)*ZEDD
   53 CONTINUE
      KORD = 0
      PA1=0.0
      PA2=0.0
      PA3=0.0
      PA4=0.0
      PA5=0.0
      PA6=0.0
      PA7=0.0
      PA8=0.0
      PA9=0.0
      PM1=0.0
      PM2=0.0
      PM3=0.0
      PM4=0.0
      PM5=0.0
      PM6=0.0
      PM7=0.0
      PM8=0.0
      PM9=0.0
      PB1=0.0
      PB2=0.0
      PB3=0.0
      PB4=0.0
      PB5=0.0
```

```
      PB6=0.0
      PB7=0.0
      PB8=0.0
      PB9=0.0
      PN1=0.0
      PN2=0.0
      PN3=0.0
      PN4=0.0
      PN5=0.0
      PN6=0.0
      PN7=0.0
      PN8=0.0
      PN9=0.0
      DO 6 I=1,NOCC
      DO 6 J=NAM,N
      IF(NAT(IA).GT.10) GO TO 54
      QAX = C(I,LY)*C(J,LZ)-C(I,LZ)*C(J,LY)
      QAY = C(I,LX)*C(J,LZ)-C(I,LZ)*C(J,LX)
      QAZ = C(I,LX)*C(J,LY)-C(I,LY)*C(J,LX)
      QDX = 0.0
      QDY = 0.0
      QDZ = 0.0
      GO TO 55
   54 CONTINUE
      QAX = C(I,LY)*C(J,LZ)-C(I,LZ)*C(J,LY)
      QAY = C(I,LX)*C(J,LZ)-C(I,LZ)*C(J,LX)
      QAZ = C(I,LX)*C(J,LY)-C(I,LY)*C(J,LX)
      QDX = C(I,LDC)*C(J,LDD)-C(I,LDD)*C(J,LDC)
     1     +C(I,LDE)*C(J,LDB)-C(I,LDB)*C(J,LDE)
     2     +(C(I,LDC)*C(J,LDA)-C(I,LDA)*C(J,LDC))*1.732
      QDY = C(I,LDB)*C(J,LDD)-C(I,LDD)*C(J,LDB)
     1     +C(I,LDC)*C(J,LDE)-C(I,LDE)*C(J,LDC)
     2     +(C(I,LDA)*C(J,LDB)-C(I,LDB)*C(J,LDA))*1.732
      QDZ = C(I,LDB)*C(J,LDC)-C(I,LDC)*C(J,LDB)
     1     +(C(I,LDD)*C(J,LDE)-C(I,LDE)*C(J,LDD))*2.000
   55 CONTINUE
      KB = 1
      SUMX=0.0
      SUMY=0.0
      SUMZ=0.0
      SDMX = 0.0
      SDMY = 0.0
      SDMZ = 0.0
      DO 5 IB=1,NA
      IF(NAT(IB)-1)4,3,4
    3 KB = KB+1
      GO TO 5
    4 KX = KB+1
      KY = KB+2
      KZ = KB+3
      IF(NAT(IB).GT.10) GO TO 56
      KB = KB+4
      GO TO 57
   56 CONTINUE
      KDA = KB+4
      KDB = KB+5
      KDC = KB+6
      KDD = KB+7
      KDE = KB+8
```

```
      KB = KB+9
      SUMX = SUMX + C(I,KY)*C(J,KZ)-C(I,KZ)*C(J,KY)
      SUMY = SUMY + C(I,KX)*C(J,KZ)-C(I,KZ)*C(J,KX)
      SUMZ = SUMZ + C(I,KX)*C(J,KY)-C(I,KY)*C(J,KX)
      SDMX = SDMX + C(I,KDC)*C(J,KDD)-C(I,KDD)*C(J,KDC)
     1              +C(I,KDE)*C(J,KDB)-C(I,KDB)*C(J,KDE)
     2              +(C(I,KDC)*C(J,KDA)-C(I,KDA)*C(J,KDC))*1.732
      SDMY = SDMY + C(I,KDB)*C(J,KDD)-C(I,KDD)*C(J,KDB)
     1              +C(I,KDC)*C(J,KDE)-C(I,KDE)*C(J,KDC)
     2              +(C(I,KDA)*C(J,KDB)-C(I,KDB)*C(J,KDA))*1.732
      SDMZ = SDMZ + C(I,KDB)*C(J,KDC)-C(I,KDC)*C(J,KDB)
     1              +(C(I,KDD)*C(J,KDE)-C(I,KDE)*C(J,KDD))*2.000
      GO TO 5
   57 CONTINUE
      SUMX = SUMX+C(I,KY)*C(J,KZ)-C(I,KZ)*C(J,KY)
      SUMY = SUMY+C(I,KX)*C(J,KZ)-C(I,KZ)*C(J,KX)
      SUMZ = SUMZ+C(I,KX)*C(J,KY)-C(I,KY)*C(J,KX)
      SDMX = SDMX + 0.0
      SDMY = SDMY + 0.0
      SDMZ = SDMZ + 0.0
    5 CONTINUE
      QAY =-QAY
      SUMY =-SUMY
      KORD = KORD + 1
      EST = CIM(KORD)
      PA1 = PA1 + QAX*SUMX/EST
      PA2 = PA2 + QAX*SUMY/EST
      PA3 = PA3 + QAX*SUMZ/EST
      PA4 = PA4 + QAY*SUMX/EST
      PA5 = PA5 + QAY*SUMY/EST
      PA6 = PA6 + QAY*SUMZ/EST
      PA7 = PA7 + QAZ*SUMX/EST
      PA8 = PA8 + QAZ*SUMY/EST
      PA9 = PA9 + QAZ*SUMZ/EST
      PM1 = PM1 + QDX*SDMX/EST
      PM2 = PM2 + QDX*SDMY/EST
      PM3 = PM3 + QDX*SDMZ/EST
      PM4 = PM4 + QDY*SDMX/EST
      PM5 = PM5 + QDY*SDMY/EST
      PM6 = PM6 + QDY*SDMZ/EST
      PM7 = PM7 + QDZ*SDMX/EST
      PM8 = PM8 + QDZ*SDMY/EST
      PM9 = PM9 + QDZ*SDMZ/EST
      EIJ = (AII(J) - AII(I)) * 27.204
      PB1 = PB1 + QAX*SUMX/EIJ
      PB2 = PB2 + QAX*SUMY/EIJ
      PB3 = PB3 + QAX*SUMZ/EIJ
      PB4 = PB4 + QAY*SUMX/EIJ
      PB5 = PB5 + QAY*SUMY/EIJ
      PB6 = PB6 + QAY*SUMZ/EIJ
      PB7 = PB7 + QAZ*SUMX/EIJ
      PB8 = PB8 + QAZ*SUMY/EIJ
      PB9 = PB9 + QAZ*SUMZ/EIJ
      PN1 = PN1 + QDX*SDMX/EIJ
      PN2 = PN2 + QDX*SDMY/EIJ
      PN3 = PN3 + QDX*SDMZ/EIJ
      PN4 = PN4 + QDY*SDMX/EIJ
      PN5 = PN5 + QDY*SDMY/EIJ
      PN6 = PN6 + QDY*SDMZ/EIJ
```

```
        PN7 = PN7 + QDZ*SDMX/EIJ
        PN8 = PN8 + QDZ*SDMY/EIJ
        PN9 = PN9 + QDZ*SDMZ/EIJ
        IF(JATOM.NE.1) GO TO 6
        IF(NAT(IA).GT.10) GO TO 58
        SXX = QAX * SUMX
        SYY = QAY * SUMY
        SZZ = QAZ * SUMZ
        RSL = ZSLTR(IA)*ZSLTR(IA)*ZSLTR(IA)/24.0
        GO TO 59
  58    CONTINUE
        SXX = (QAX+QDX)*(SUMX+SDMX)
        SYY = (QAY+QDY)*(SUMY+SDMY)
        SZZ = (QAZ+QDZ)*(SUMZ+SDMZ)
        RSL = ZSLTR(IA)*ZSLTR(IA)*ZSLTR(IA)/81.0
  59    CONTINUE
        AXX = -(SXX/EST) * 2.14694 * 3.0 * RSL * 449.903
        AYY = -(SYY/EST) * 2.14694 * 3.0 * RSL * 449.903
        AZZ = -(SZZ/EST) * 2.14694 * 3.0 * RSL * 449.903
        BXX = -(SXX/EIJ) *2.14694 * 3.0 * RSL * 449.903
        BYY = -(SYY/EIJ) *2.14694 * 3.0 * RSL * 449.903
        BZZ = -(SZZ/EIJ) *2.14694 * 3.0 * RSL * 449.903
        WRITE(IY,113) I,J,KORD,EIJ,EST,SXX,SYY,SZZ,
       1 AXX,AYY,AZZ,BXX,BYY,BZZ
 113    FORMAT(I3,1X,I3,I4,4X,F6.3,4X,F6.3,3F10.4,6F10.3)
        ITT=0
        IF((ITYPE(I).EQ.IS).AND.(ITYPE(J).EQ.IS))ITT=1
        IF((ITYPE(I).EQ.IS).AND.(ITYPE(J).EQ.IP))ITT=2
        IF((ITYPE(I).EQ.IP).AND.(ITYPE(J).EQ.IS))ITT=3
        IF((ITYPE(I).EQ.IN).AND.(ITYPE(J).EQ.IS))ITT=4
        IF((ITYPE(I).EQ.IN).AND.(ITYPE(J).EQ.IP))ITT=5
        CALL TRANS(ITT,1,AXX,AYY,AZZ,BXX,BYY,BZZ)
   6    CONTINUE
        IF(JATOM.NE.1) GO TO 700
        WRITE(6,500)
 500    FORMAT(///28X,10HEJ-EI+2K-J,41X,10H   EJ - EI ,/2X,11HTRANSITIONS,
       1 17X,1HX,11X,1HY,11X,1HZ,9X,7HAVERAGE,10X,1HX,11X,1HY,
       2 11X,1HZ,9X,7HAVERAGE/)
        DO 600 IT1=1,5
        WRITE(6,610) TRA(IT1),(XTOY(IT1,IT2),IT2=1,8)
 610    FORMAT(/3X,A8,3X,4(2X,F10.3),3X,4(2X,F10.3))
 600    CONTINUE
 700    CONTINUE
        GAMMA(IA,1) = PA1*2.14694
        GAMMA(IA,2) = PA2*2.14694
        GAMMA(IA,3) = PA3*2.14694
        GAMMA(IA,4) = PA4*2.14694
        GAMMA(IA,5) = PA5*2.14694
        GAMMA(IA,6) = PA6*2.14694
        GAMMA(IA,7) = PA7*2.14694
        GAMMA(IA,8) = PA8*2.14694
        GAMMA(IA,9) = PA9*2.14694
        GAMMB(IA,1) = PM1*2.14694
        GAMMB(IA,2) = PM2*2.14694
        GAMMB(IA,3) = PM3*2.14694
        GAMMB(IA,4) = PM4*2.14694
        GAMMB(IA,5) = PM5*2.14694
        GAMMB(IA,6) = PM6*2.14694
        GAMMB(IA,7) = PM7*2.14694
```

5 PROGRAM LISTINGS

```
      GAMMB(IA,8) = PM8*2.14694
      GAMMB(IA,9) = PM9*2.14694
      GGG(IA,1) = PB1*2.14694
      GGG(IA,2) = PB2*2.14694
      GGG(IA,3) = PB3*2.14694
      GGG(IA,4) = PB4*2.14694
      GGG(IA,5) = PB5*2.14694
      GGG(IA,6) = PB6*2.14694
      GGG(IA,7) = PB7*2.14694
      GGG(IA,8) = PB8*2.14694
      GGG(IA,9) = PB9*2.14694
      GGGG(IA,1) = PN1*2.14694
      GGGG(IA,2) = PN2*2.14694
      GGGG(IA,3) = PN3*2.14694
      GGGG(IA,4) = PN4*2.14694
      GGGG(IA,5) = PN5*2.14694
      GGGG(IA,6) = PN6*2.14694
      GGGG(IA,7) = PN7*2.14694
      GGGG(IA,8) = PN8*2.14694
      GGGG(IA,9) = PN9*2.14694
    7 CONTINUE
      WRITE(IY,107)
      WRITE(IY,100)
      WRITE(IY,101)
      DO 19 IA = 1,NA
      IF (NAT(IA)-1) 18,19,18
   18 IF(NAT(IA).GT.10) GO TO 181
      DIAA = 17.7503*(2.0*(NAT(IA)-0.3) + P(IA,IA)*ZN(IA)/2.0)
      GO TO 60
  181 DIAA = 17.7503*(4.0*(NAT(IA)-2.225) + P(IA,IA)*ZN(IA)/3.0)
      GO TO 61
   60 IB = NAT(IA)-4
      IBB = NAT(IA)-4
      Q1 = Q(IA)
      ZES = ZSLTR(IA)
      ZEB = ZP(IA)
      ZEFFS = ZES*ZES*ZES
      ZEFFB = ZEB*ZEB*ZEB
      RR = ZEFFS/24.0
      RS(IA) = RR
      RRB = ZEFFB/24.0
      RB(IA) = RRB
      RN = 0.0
      RD(IA) = 0.0
      GO TO 62
   61 IB = NAT(IA) - 7
      IBB = NAT(IA) - 7
      ZES = ZSLTR(IA)
      ZEB = 0.0
      ZEFFS = ZES*ZES*ZES
      ZEFFB = 0.0
      RR = ZEFFS/81.0
      RS(IA) = RR
      RN = ZEFFS/405.0
      RD(IA) = RN
      RRB = 0.0
      RB(IA) = 0.0
   62 CONTINUE
      PARA = -((GAMMA(IA,1) + GAMMA(IA,5) + GAMMA(IA,9))*RR
```

THEORY OF NMR PARAMETERS

```
     1         + (GAMMB(IA,1) + GAMMB(IA,5) + GAMMB(IA,9))*RN)
     2            *449.903
      PARAS(IA) = PARA
      PARAB(IA) = -((GAMMA(IA,1) + GAMMA(IA,5) + GAMMA(IA,9))*RRB
     1         + (GAMMB(IA,1) + GAMMB(IA,5) + GAMMB(IA,9))*RRB)
     2            *449.903
      PARASE(IA) = -((GGG(IA,1) + GGG(IA,5) + GGG(IA,9))*RR
     1         + (GGGG(IA,1) + GGGG(IA,5) + GGGG(IA,9))*RN)
     2            *449.903
      PARABE(IA) = -((GGG(IA,1) + GGG(IA,5) + GGG(IA,9))*RRB
     1         + (GGGG(IA,1) + GGGG(IA,5) + GGGG(IA,9))*RRB)
     2            *449.903
      PARCE1 = 0.0
      SIGA = 0.0
      IF(NAT(IA).GT.10) GO TO 63
      IF(Q1.GT.0.0) SIGA = DIA(2,IB)-Q1*(DIA(2,IB)-DIA(1,IB))
      IF(Q1.LT.0.0) SIGA = DIA(2,IB)-Q1*(DIA(3,IB)-DIA(2,IB))
  63 CONTINUE
      DIAC = 0.0
      PARC = 0.0
      SIGC = 0.0
      NEN = 0
      DO 20 IC = 1,NA
      IF (R(IC,IA) - 2.5) 21,20,20
  21 IF (R(IC,IA) - 0.1) 20,22,22
  22 NEN = NEN + 1
      IF(NAT(IC).GT.10) GO TO 66
      SIGC = SIGC + 9.3929 * (NAT(IC)-Q(IC))/R(IC,IA)
      GO TO 67
  66 SIGC = 0.0
  67 CONTINUE
      IF (NAT(IC) - 1) 23,20,23
  23 RC2 = R(IC,IA)*R(IC,IA)
      RC5 = RC2*RC2*R(IC,IA)
      RX = ABS(XC(IC)-XC(IA))
      RY = ABS(YC(IC)-YC(IA))
      RZ = ABS(ZC(IC)-ZC(IA))
      RXX = 3.0*RX*RX
      RXY = 3.0*RX*RY/RC5
      RXZ = 3.0*RX*RZ/RC5
      RYY = 3.0*RY*RY
      RYZ = 3.0*RY*RZ/RC5
      RZZ = 3.0*RZ*RZ
      XR=(RC2-RXX)/RC5
      YR = (RC2 - RYY)/RC5
      ZR = (RC2 - RZZ)/RC5
      DIAC = DIAC - (BETA(IC,1)*XR+BETA(IC,2)*YR+BETA(IC,3)*ZR
     1         +BETA(IC,4)*XR+BETA(IC,5)*YR+BETA(IC,6)*ZR)/3.0
      PARC = PARC - (GAMMA(IC,1)*XR - GAMMA(IC,2)*RXY - GAMMA(IC,3)*RXZ
     1            - GAMMA(IC,4)*RXY + GAMMA(IC,5)*YR - GAMMA(IC,6)*RYZ
     2            - GAMMA(IC,7)*RXZ - GAMMA(IC,8)*RYZ + GAMMA(IC,9)*ZR
     3            + GAMMB(IC,1)*XR - GAMMB(IC,2)*RXY - GAMMB(IC,3)*RXZ
     4            - GAMMB(IC,4)*RXY + GAMMB(IC,5)*YR - GAMMB(IC,6)*RYZ
     5            - GAMMB(IC,7)*RXZ - GAMMB(IC,8)*RYZ + GAMMB(IC,9)*ZR)
     6            *33.339
      PARCE1=PARCE1 - (GGG(IC,1)*XR - GGG(IC,2)*RXY - GGG(IC,3)*RXZ
     1            - GGG(IC,4)*RXY + GGG(IC,5)*YR - GGG(IC,6)*RYZ
     2            - GGG(IC,7)*RXZ - GGG(IC,8)*RYZ + GGG(IC,9)*ZR
     3            + GGGG(IC,1)*XR - GGGG(IC,2)*RXY - GGGG(IC,3)*RXZ
```

5 PROGRAM LISTINGS

```
     4                  - GGGG(IC,4)*RXY + GGGG(IC,5)*YR - GGGG(IC,6)*RYZ
     5                  - GGGG(IC,7)*RXZ - GGGG(IC,8)*RYZ + GGGG(IC,9)*ZR)
     6                  *33.339
  20 CONTINUE
     IF(NAT(IA).LT.6) SIGC = 0.0
     DIACC(IA) = DIAC
     PARCC(IA) = PARC
     PARCE(IA) = PARCE1
     TOTP = DIAA + PARA + DIAC + PARC
     IF(NAT(IA).GT.10) GO TO 64
     TOTF = SIGA + SIGC
     GO TO 65
  64 TOTF = 0.0
  65 CONTINUE
     WRITE(IY,103) ATOM(IBB),IA,RR,NEN,DIAA,SIGA,PARA,DIAC,SIGC,PARC,
    1 TOTP,TOTF
  19 CONTINUE
     DO 444 ID=1,NA
     IF(NAT(ID).EQ.1) GO TO 444
     PCONT = (GAMMA(ID,1)+GAMMA(ID,5)+GAMMA(ID,9))*449.903
     DCONT = (GAMMB(ID,1)+GAMMB(ID,5)+GAMMB(ID,9))*449.903
     WRITE(6,445) PCONT,ID,RS(ID),DCONT,ID,RD(ID)
 445 FORMAT(10X,6HPCONT=,F11.5,5X,3HRS(,I2,2H)=,F6.3,
    1       /10X,6HDCONT=,F11.5,5X,3HRD(,I2,2H)=,F6.3)
 444 CONTINUE
     WRITE(IY,120)
     WRITE(IY,121)
     WRITE(IY,122)
     WRITE(IY,123)
     DO 31 IA=1,NA
     IF(NAT(IA)-1) 30,31,30
  30 IF(NAT(IA).GT.10) GO TO 80
     IB = NAT(IA) - 4
     GO TO 81
  80 IB = NAT(IA) - 7
  81 CONTINUE
     TOTS = DIAS(IA) + DIACC(IA) + PARAS(IA) + PARCC(IA)
     TOTB = DIAB(IA) + DIACC(IA) + PARAB(IA) + PARCC(IA)
     WRITE(6,124) ATOM(IB),IA,RS(IA),DIAS(IA),DIACC(IA),PARAS(IA),
    1 PARCC(IA),TOTS,RB(IA),DIAB(IA),DIACC(IA),PARAB(IA),
    2 PARCC(IA),TOTB
  31 CONTINUE
     WRITE(IY,125)
     WRITE(IY,121)
     WRITE(IY,126)
     WRITE(IY,127)
     DO 33 IA=1,NA
     IF(NAT(IA)-1) 32,33,32
  32 IF(NAT(IA).GT.10) GO TO 86
     IB = NAT(IA) - 4
     GO TO 83
  86 IB = NAT(IA) - 7
  83 CONTINUE
     TOTS = DIAS(IA) + DIACC(IA) + PARASE(IA) + PARCE(IA)
     TOTB = DIAB(IA) + DIACC(IA) + PARABE(IA) + PARCE(IA)
     WRITE(6,128) ATOM(IB),IA,ZSLTR(IA),DIAS(IA),DIACC(IA),
    1 PARASE(IA),PARCE(IA),TOTS,ZS(IA),ZP(IA),DIAB(IA),DIACC(IA),
    2 PARABE(IA),PARCE(IA),TOTB
  33 CONTINUE
```

```
      WRITE(IY,129)
      WRITE(IY,130)
      WRITE(IY,131)
      WRITE(IY,132)
      WRITE(IY,133)
      WRITE(IY,134)
      DO 35 IA=1,NA
      IF(NAT(IA)-1) 34,35,34
   34 IF(NAT(IA).GT.10) GO TO 84
      IB = NAT(IA) - 4
      GO TO 85
   84 IB = NAT(IA) - 7
   85 CONTINUE
      Q1 = Q(IA)
      CHG = P(IA,IA)
      ITN = 0
      DO 800 IT1=1,3
      DO 800 IT2=1,3
      ITN=ITN+1
      GT1(IT1,IT2) = -(GAMMA(IA,ITN)*RS(IA) + GAMMB(IA,ITN)*RD(IA))
     1                *3.0*449.903
      GT2(IT1,IT2) = -(GGG(IA,ITN)*RS(IA) + GGGG(IA,ITN)*RD(IA))
     1                *3.0*449.903
  800 CONTINUE
      WRITE(6,135) ATOM(IB),IA,CHG,Q1,(GT1(1,I2),I2=1,3),
     1 (GT2(1,I2),I2=1,3)
      DO 6000 I1=2,3
 6000 WRITE(6,140) (GT1(I1,I2),I2=1,3),(GT2(I1,I2),I2=1,3)
      NN = 3
      L = 1
      CALL F02AFF(GT1,3,NN,ER,EI,ITS,L)
      IF (L.NE.0) GO TO 3000
      WRITE(IY,1000)
 1000 FORMAT(///31X,30H DIAGONALIZED SHIELDING TENSOR)
      WRITE(IY,2000) (ER(I),EI(I),I=1,NN)
 2000 FORMAT(/33X,1X,2E20.6)
      GO TO 8000
 3000 WRITE(IY,7000)
 7000 FORMAT(/,1X,23HMORE THAN 30 ITERATIONS)
 8000 CONTINUE
   35 CONTINUE
      CALL AEE
  100 FORMAT (//5H ATOM,3X,4HATOM,5X,4HINV.,4X,2HNN,6X,4HDIAA,7X,
     1 4HDIAA,7X,4HPARA,8X,4HDIAC,6X,4HDIAC,7X,4HPARC,8X,3HTOT,7X,3HTOT)
  101 FORMAT(1H ,8X,3HNO.,6X,2HR3,14X,3HPOP,8X,3HFLY,8X,3HPOP,9X,
     1 3HPOP,7X,3HFLY,8X,3HPOP,8X,3HPOP,7X,3HFLY)
  103 FORMAT (/A8,I3,4X,F6.3,4X,I2,8F11.2)
  107 FORMAT(1X,40H SHIFTS BEFORE CONFIGURATION INTERACTION)
  120 FORMAT(////30X,41H THE DIAMAGNETIC TERM IS EVALUATED BY THE,
     1 23H USUAL POPLE EXPRESSION/)
  121 FORMAT(//37H EFFECTIVE CHARGE Z FROM SLATER RULES,33X,
     1 36H EFFECTIVE CHARGE Z FROM BURNS RULES/)
  122 FORMAT(/5H ATOM,3X,4HATOM,2X,4HINV.,4X,4HDIAA,4X,4HDIAC,
     1 4X,4HPARA,4X,4HPARC,4X,4HTOTS,12X,4HINV.,4X,4HDIAA,4X,
     2 4HDIAC,4X,4HPARA,4X,4HPARC,4X,4HTOTB)
  123 FORMAT(1H ,8X,3HNO.,3X,2HR3,5X,4H POP,4X,4H POP,4X,4H POP,
     1 4X,4H POP,4X,4H POP,12X,4H R3 ,4X,4H POP,4X,4H POP,4X,4H POP,
     2 4X,4H POP,4X,4H POP)
  124 FORMAT(/A8,I3,F7.3,5F8.2,10X,F6.3,5F8.2)
```

```
    125 FORMAT(////48H TRANSITION ENERGIES ARE TAKEN AS THE DIFFERENCE,
       1 57H IN ORBITAL ENERGIES (EJ - EI) WITHOUT COULOMB INTEGRALS /)
    126 FORMAT(/5H ATOM,3X,4HATOM,6H   ZE   ,2X,6H DIAA ,4X,4HDIAC,
       1 4X,4HPARA,4X,4HPARC,4X,4HTOTS,6X,6H Z2S    ,6H Z2P    ,4X,
       2 4HDIAA,4X,4HDIAC,4X,4HPARA,4X,4HPARC,4X,4HTOTB)
    127 FORMAT(1H ,8X,3HNO.,7X,6H    POP,5X,4H POP,
       1 4X,4H POP,4X,4H POP,4X,4H POP,6X,12HBURNS RULES ,4X,4H POP,
       2 4X,4H POP,4X,4H POP,4X,4H POP,4X,4H POP)
    128 FORMAT(/A8,I3,F7.4,5F8.2,4X,F6.4,2X,F6.4,5F8.2)
    129 FORMAT(////40X,36H PARAMAGNETIC SHIELDING ANISOTROPIES/)
    130 FORMAT(/37X,42H THE INV. R3 FACTOR IS FROM SLATER RULES   /)
    131 FORMAT(/38X,20H EJ - EI + 2K - J   ,14X,20H     EJ - EI        )
    132 FORMAT (38X,20HSHIELDING COMPONENTS,14X,20HSHIELDING COMPONENTS)
    133 FORMAT(/5H ATOM,3X,4HATOM,3X,6H TOTAL,3X,6H   NET ,8X,4H XX  ,4X,
       1 4H YY  ,4X,4H ZZ  ,14X,4H XX  ,4X,4H YY  ,4X,4H ZZ  )
    134 FORMAT(1H ,8X,3HNO.,3X,6HCHARGE,3X,6HCHARGE)
    135 FORMAT(/A8,I3,3X,F7.4,2X,F7.4,5X,3F8.2,10X,3F8.2)
    140 FORMAT(35X,3F8.2,10X,3F8.2)
        RETURN
        END

        SUBROUTINE AEE
C       AVERAGE EXCITATION ENERGY CALCULATIONS
        COMMON/ONE/ N,NA,NOCC,ICHGE,IX,IY,I,J,KI,KJ,ISUB,IDUMB,ISTOP,
       1 E,TCONV,MU,MO
        COMMON/THREE/ NAT(35),S(15),ZN(35),ZCHG(35),Q(35)
        COMMON/FOUR/ P(35,35),QAB(35,35)
        COMMON/FIVE/ PO(100,100),XXX(525),ZSLTR(35),ZS(35),ZP(35),
       1              YYY(105),RS(35),RB(35),QAA(35)
        COMMON/SIX/ U(35,3),R(35,35),BETA(35,35),GAMMA(35,35)
        DIMENSION ATOM(10)
        DATA ATOM(1)/8H    B   /
        DATA ATOM(2)/8H    C   /
        DATA ATOM(3)/8H    N   /
        DATA ATOM(4)/8H    O   /
        DATA ATOM(5)/8H    F   /
        DATA ATOM(6)/8H    AL  /
        DATA ATOM(7)/8H    SI  /
        DATA ATOM(8)/8H    P   /
        DATA ATOM(9)/8H    S   /
        DATA ATOM(10)/8H    CL  /
        WRITE(6,100)
        WRITE(6,200)
        WRITE(6,300)
        WRITE(6,400)
        KQ = 1
        DO 11 IA=1,NA
        IF(NAT(IA)-1) 2,1,2
   1    KQ = KQ + 1
        GO TO 11
   2    KS = KQ
        KX = KQ + 1
        KY = KQ + 2
        KZ = KQ + 3
        IF(NAT(IA).LT.10) GO TO 60
        KDA = KQ+4
        KDB = KQ+5
        KDC = KQ+6
        KDD = KQ+7
```

```
          KDE = KQ+8
          KQ = KQ+9
          GO TO 61
       60 KQ = KQ+4
          QP = P(IA,IA) - PO(KS,KS)
          QAA(IA) = (4.0/3.0)*(QP - 0.5*(PO(KX,KX)*PO(KY,KY) + PO(KX,KX)*
         1 PO(KZ,KZ) + PO(KY,KY)*PO(KZ,KZ)) + 0.5*(PO(KX,KY)*PO(KX,KY) +
         2 PO(KX,KZ)*PO(KX,KZ) + PO(KY,KZ)*PO(KY,KZ)))
          GO TO 68
       61 QP = 0.0
          QAA(IA) = 0.0
       68 CONTINUE
          IN = 0
          SUMQAB = 0.0
          LQ = 1
          DO 8 IC=1,NA
          IF(NAT(IC) - 1) 4,3,4
        3 LQ = LQ + 1
          GO TO 8
        4 LS = LQ
          LX = LQ + 1
          LY = LQ + 2
          LZ = LQ + 3
          IF(NAT(IC).LT.10) GO TO 62
          LDA = LQ+4
          LDB = LQ+5
          LDC = LQ+6
          LDD = LQ+7
          LDE = LQ+8
          LQ = LQ+9
          GO TO 66
       62 LQ = LQ + 4
       66 CONTINUE
          IF(IC.EQ.IA) GO TO 8
          IF(NAT(IA).GT.10) GO TO 63
          IN = IN + 1
          QAB(IA,IN) = (2.0/3.0)*( -(PO(KX,LX)*PO(KY,LY) + PO(KX,LX)*
         1 PO(KZ,LZ) + PO(KY,LY)*PO(KZ,LZ)) + (PO(KX,LY)*PO(LX,KY)
         2 + PO(KX,LZ)*PO(LX,KZ) + PO(KY,LZ)*PO(LY,KZ)))
          GO TO 64
       63 QAB(IA,IN) = 0.0
       64 CONTINUE
          SUMQAB = SUMQAB + QAB(IA,IN)
        8 CONTINUE
          IF(NAT(IA).GT.10) GO TO 65
          SUMQAB = SUMQAB + QAA(IA)
          ZES = ZSLTR(IA)
          ZEB = ZP(IA)
          GO TO 67
       65 CONTINUE
          SUMQAB = 0.0
          ZES = ZSLTR(IA)
          ZEB = ZP(IA)
       67 CONTINUE
          TERMS = ZES*ZES*ZES * SUMQAB
          TERMB = ZEB*ZEB*ZEB * SUMQAB
          PARAS = - SUMQAB * RS(IA) * 724.43605
          PARAB = - SUMQAB * RB(IA) * 724.43605
       10 IF(NAT(IA).GT.10) GO TO 70
```

```
      IB = NAT(IA) - 4
      GO TO 71
 70   IB = NAT(IA) - 7
 71   CONTINUE
      WRITE(6,500) ATOM(IB),IA,QAA(IA),SUMQAB,
     1 TERMS,TERMB,PARAS,PARAB
 11   CONTINUE
      WRITE(6,600)
600   FORMAT(///20X,18H CHARGE DENSITIES /
     1 5H ATOM,3X,4HATOM,5X,7H   S    ,7H   PX   ,7H   PY   ,
     2 7H   PZ   ,7H DZ**2 ,7H  DXZ   ,7H  DYZ   ,7HD(X-Y) ,
     3 7H  DXY   ,7H QTOTAL/9X,3HNO.,5X,10(1X,5(1H*),1X)/)
      KR =1
      DO 50 IA=1,NA
      IF(NAT(IA).EQ.1) GO TO 30
      IF(NAT(IA).GT.10) GO TO 40
      IB = NAT(IA) - 4
      KR2 = KR + 3
      WRITE(6,610) ATOM(IB),IA,(PO(KF,KF),KF=KR,KR2),P(IA,IA)
      KR = KR + 4
      GO TO 50
 40   IB = NAT(IA) - 7
      KR2 = KR + 8
      WRITE(6,640) ATOM(IB),IA,(PO(KF,KF),KF=KR,KR2),P(IA,IA)
      KR = KR + 9
      GO TO 50
 30   WRITE(6,620) IA,PO(KR,KR),P(IA,IA)
      KR = KR + 1
 50   CONTINUE
      WRITE(6,630)
610   FORMAT(/A8,I3,6X,4F7.4,35X,F7.4)
640   FORMAT(/A8,I3,6X,10F7.4)
620   FORMAT(/8H    H    ,I3,6X,F7.4,56X,F7.4)
630   FORMAT(//20X,70(1H*)/)
100   FORMAT(////40X,41H AVERAGE EXCITATION ENERGY APPROXIMATION )
200   FORMAT(/45X,30H PARA = -724.436*(R-3)*SUMQAB /)
300   FORMAT(//5H ATOM,3X,4HATOM,2X,8H   QAA    ,44X,8H SUMQAB  ,4X,
     2 5X,10H Z3*SUMQAB,5X,4X,8H   PARA   ,2X,8H   PARA   )
400   FORMAT(1H ,8X,3HNO.,66X,8H SLATER  ,4X,8H  BURNS  ,4X,8H SLATER ,
     1 2X,8H  BURNS  )
500   FORMAT(/A8,I3,3X,F8.5,44X,F8.5,4X,F8.4,4X,F8.4,4X,F8.2,2X,F8.2)
      RETURN
      END

      SUBROUTINE PROUT(X,Y,M,NDIM)
      DIMENSION X(NDIM,NDIM),Y(NDIM,NDIM)
      DIMENSION Z(5)
      I = -4
 1    I = I+5
      IF(I-M+3)6,2,2
 6    I4 = I+4
      GO TO 3
 2    IF(I-M)7,7,5
 7    I4 = M
 3    WRITE (6,1000)(X(K,K),Y(K,1),K=I,I4)
      DO 4 J=2,M
      L = 0
      DO 40 II=I,I4
      L = L + 1
```

```
      Z(L) = Y(II,J)
   40 CONTINUE
    4 WRITE(6,1001)(Z(K),K=1,L)
      GO TO 1
    5 CONTINUE
 1000 FORMAT (/2F10.6,4(F15.6,F10.6))
 1001 FORMAT (F20.6,4F25.6)
      RETURN
      END

      SUBROUTINE MATRIX (X,E,N,NDIM,KNTRL)
      DIMENSION X(NDIM,NDIM),E(NDIM,NDIM)
      NSQP=3*N*N/2
      ITER=0
      IF(N-1)187,2,3
    2 E(1,1)=1.
      GO TO 187
    3 IF(KNTRL)7,9,7
    9 DO 11 I=1,N
      DO 10 J=I,N
      X(J,I)=X(I,J)
      E(I,J) = 0.0
   10 E(J,I) = 0.0
   11 E(I,I) =1.0
      DO 12 K=1,N
   12 E(K,K)=1.
    7 ITER=ITER+1
   21 K=1
      L=2
   22 BIGX=ABS(X(2,1))
      DO 26 J=3,N
      JM1=J-1
      DO 25 I=1,JM1
      DIFR=BIGX-ABS(X(I,J))
      IF (DIFR) 23,25,25
   23 BIGX=ABS(X(I,J))
      K=I
      L=J
   25 CONTINUE
   26 CONTINUE
      IF (BIGX-0.00001) 76,76,27
   27 TS=X(K,L)*X(K,L)
   28 DEL=X(K,K)-X(L,L)
      R=SQRT(ABS(DEL*DEL+4.0*TS))
      A=SQRT(ABS((R+DEL)/(2.0*R)))
      IF(.707-A) 30,30,29
   29 B=0.-A
      A=SQRT(1.0-B*B)
      GO TO 32
   30 B=-SQRT(1.0-A*A)
   32 IF (DEL/X(K,L)) 33,60,60
   33 B=0.-B
   60 DO 65 J=1,N
      IF (K-J) 61,65,61
   61 IF (L-J) 62,65,62
   62 X(K,J)=A*X(J,K)-B*X(J,L)
      X(L,J)=B*X(J,K)+A*X(J,L)
   65 CONTINUE
      DO 70 J=1,N
```

```
   67 X(J,K)=X(K,J)
      X(J,L)=X(L,J)
   70 CONTINUE
      D=X(K,K)+X(L,L)
      X(K,K)=A*A*X(K,K)+B*B*X(L,L)-2.*A*B*X(K,L)
      X(L,L)=D-X(K,K)
      X(L,K)=0.
      X(K,L)=0.
   71 DO 75 J=1,N
      F1=E(K,J)
      E(K,J)=A*F1-B*E(L,J)
   75 E(L,J)=B*F1+A*E(L,J)
      IF (ITER-NSQP) 7,90,90
   90 ITER=0
      WRITE(6,91)
   91 FORMAT (26H TOUGH MATRIX SKIPPED OUT )
      GO TO 87
   76 BIGR=0.00004
      NMI=N-1
      DO 86 I=1,NMI
      IPI=I+1
      DO 86 J=IPI,N
      IF(X(I,J))77,86,77
   77 DEL=X(I,I)-X(J,J)
      IF (DEL) 79,86,79
   79 ROT=ABS(X(I,J)/DEL)
      IF (ROT-BIGR) 86,86,78
   78 BIGR=ROT
      K=I
      L=J
   86 CONTINUE
      IF (BIGR-0.00005) 87,87,27
   87 NEWN=N-1
      DO 52 L=1,NEWN
      J=L
      I=L+1
   92 IF(X(J,J)-X(I,I)) 93,100,100
  100 K=I
      J=I
      IF (I-N) 101,94,101
  101 I = I+1
      GO TO 92
   93 K = J
      IF (I-N) 102,94,102
  102 I = I+1
      GO TO 92
   94 B = X(L,L)
      X(L,L) = X(K,K)
      X(K,K) = B
      DO 51 I=1,N
      B=E(K,I)
      E(K,I)=E(L,I)
   51 E(L,I)=B
   52 CONTINUE
  187 RETURN
      END
```

REFERENCES

1. M. Kondo, I. Ando, R. Chujo and A. Nishioka, *J. magn. Reson.*, **24**, 315 (1976).
2. J. A. Pople, D. L. Beveridge and P. A. Dobosh, *J. Chem. Phys.*, **47**, 2026 (1967).
3. J. A. Pople and D. L. Beveridge, "Approximate Molecular Orbital Theory", McGraw-Hill, New York (1970).
4. J. A. Pople, *J. Chem. Phys.*, **37**, 53 (1962).
5. J. A. Pople, *J. Chem. Phys.*, **37**, 60 (1962).
6. K. Krogh-Jespersen and M. Ratner, *J. Chem. Phys.*, **65**, 1305 (1976).
7. M. Jallali-Heravi and G. A. Webb, *J. magn. Reson.*, **32**, 429 (1978).
8. M. Jallali-Heravi and G. A. Webb, *Org. magn. Reson.*, **11**, 524 (1978).

Index

Ab initio molecular orbital calculations, 36
  of spin–spin couplings, 99
Acetone,
  shielding calculations of, 65
Acetonitrile,
  spin–spin coupling calculations of, 101, 103, 105
Acetylene,
  shielding calculations of, 65
  spin–spin coupling calculations of, 99, 101, 103
m-Aminobenzoic acid cation, 11
Ammonia,
  ab initio shielding calculations of, 55
  spin–spin coupling calculations of, 101
Angular dependence of spin–spin couplings, 106
Anisotropy of shielding tensor, 5, 16, 53, 69
Applied magnetic field B_0, 2
Asymmetry of electric field gradient, 13
Average excitation energy (AEE) approximation,
  applied to calculations of spin–spin couplings, 84, 89, 98
  applied to Pople's shielding model, 67
  applied to Ramsey's shielding model, 51
trans-Azobenzene,
  ^{15}N relaxation of, 16

Benzene,
  ring current contribution to proton shielding of, 70
  shielding calculations of, 65
  spin–spin coupling calculations of, 101, 105
Bohr magneton, 44
Bond order terms, 67
Born–Oppenheimer approximation, 22
Boron hydrides,
  some calculated boron shieldings of, 62
Brownian process, 10

Calculations giving gauge-independent shielding data, 57
Carbon shielding calculations of unsaturated molecules, 68
Carbon–hydrogen spin–spin couplings,
  FPT calculations of, 101
  SOS calculations of, 99
Carbon tetrachloride,
  ^{35}Cl line width of, 14
Charge density terms, 67
Chemical shift, 5
CNDO/2 method,
  approximations involved in, 33
  parameters used in, 34
Column vector, 32
Comparison of coupling results from various theories, 104
Computer program,
  for calculating nuclear shielding by FPT-INDO and CNDO/2 methods, 115
  description of input data for, 116
  for calculating nuclear shielding by INDO/S-SOS method, 160
  description of input data for, 159
  libraries, 4, 7
Contact contribution to spin–spin coupling, 44, 83, 84, 94, 96
Core Hamiltonian, 36
Correlation time, 8
Coulomb integral,
  in molecular orbital theory, 30, 91
  in valence bond theory, 40, 96
Cyclophane protons,
  shielding of, 70
Cyclopropane,
  spin–spin coupling calculations of, 101

Dewar structures, 39
Diamagnetic contribution to nuclear
    shielding, 49
  free atom values of, 56
  one-centre contribution to, 60
  three-centre contribution to, 60
  two-centre contribution to, 60
Dimethyl thallium nitrate,
  proton relaxation of, 17
Dinucleoside monophosphate,
  ^{31}P shielding study of, 60
Dipolar contribution to spin–spin
    couplings, 44, 83, 85, 95
Dirac delta function, 44, 109

Electric quadrupolar relaxation, 13
Electronic Hamiltonian operator, 22
Electronic spin functions, 28, 31
Electrostatic potential energy term, 41
Empirical molecular orbital methods, 32
Ethane,
  spin–spin coupling calculations of,
    99, 101, 103, 105, 107, 110
Ethylene,
  shielding calculations of, 65
  spin–spin coupling calculations of,
    99, 101, 103, 105
Exchange integral,
  in molecular orbital theory, 30, 91
  in valence bond theory, 40, 96
Experimental aspects of NMR, 3

Finite perturbation theory (FPT), 25
  applied to nuclear shielding
    calculations, 51, 54, 58
  applied to spin–spin coupling
    calculations, 92, 100
First-order perturbation terms, 20
Fluorine–hydrogen spin–spin couplings,
  $ab\ initio$ calculations of, 100
Formaldehyde, spin–spin coupling
    calculations of, 99, 101
Formalism of quantum chemistry, 21
Formamide,
  calculated shielding data on, 77

Gauge-dependent shielding
    calculations, 53
Gauge-invariant atomic orbitals, 58

Gaussian type orbitals, 36
Glycyl residues of haemoglobin,
  ^{15}N signals of, 17

Hamiltonian operator, 21
Hartree–Fock equations, 30
Hellmann–Feynmann theorem, 27, 52
Hückel theory, 32, 71
Hybridization effects on spin–spin
    couplings, 104
Hydrated formamide,
  some calculated shielding data on, 77
Hydrogen cyanide,
  shielding calculations of, 65
Hydrogen deuteride,
  spin–spin coupling of, 86
Hydrogen fluoride,
  $ab\ initio$ shielding calculations of, 55
  $ab\ initio$ spin–spin coupling
    calculations of, 100
  shielding density maps of, 79
Hydrogen–hydrogen spin–spin
    couplings,
  FPT calculations of, 101
  SOS calculations of, 99

INDO method, 34
  approximations involved in, 34
Intermolecular relaxation interactions, 11
Internal rotation of ammonium group, 11
Intramolecular relaxation interactions, 10

Karplus relation, 107, 109
Kekulé structures, 39
Kronecker delta function, 49, 67

Linear combination of atomic orbitals–
    self consistent field (LCAO-
    SCF) equations, 30
Local diamagnetic contribution to
    nuclear shielding, 59, 64, 66
Local paramagnetic contribution to
    nuclear shielding, 59, 64, 66
Long range proton–proton spin–spin
    couplings,
  calculations of, 101

Magnetic dipole relaxation, 10
Magnetic susceptibility tensor, 55
Matrix sum rule, 84
Mercury compounds,
  nuclear shielding anisotropy of, 16
Methane,
  ab initio shielding calculations of, 55
  spin–spin coupling calculations of, 99, 101, 105, 107
Methanol,
  spin–spin coupling calculations of, 105
Methylamine,
  spin–spin coupling calculations of, 103, 105
Methylene difluoride,
  spin–spin coupling calculations of, 103
Methyl fluoride,
  anisotropy of spin–spin coupling of, 102
  spin–spin coupling calculations of, 103, 105
N-Methylformamide,
  spin–spin coupling calculations of, 107, 110
Methyl isocyanide,
  spin–spin coupling calculations of, 103
MINDO/2 method, 35
MINDO/3 method, 35
Molecular orbital (MO) theory, 28
  applied to nuclear shielding calculations, 53
  applied to spin–spin coupling calculations, 88

Nitrogen molecule,
  shielding calculations of, 65
Nitromethane,
  effect of solvent on nitrogen shielding of, 79
NMR,
  general remarks on, 1
  parameters, 1
Non-empirical molecular orbital methods, 36
Non-local contributions to nuclear shielding, 59

Non-local diamagnetic contribution to nuclear shielding, 59, 64
Non-local paramagnetic contribution to nuclear shielding, 59, 64
Non-specific solvent effects on nuclear shielding, 76
Nuclear magnetic moment, 2
Nuclear shielding anisotropy, 5
  calculations of based upon SCF theories, 63
  nuclear relaxation due to, 16
Nuclear shielding, 4, 41, 47
  diamagnetic contributions to, 49, 56, 64, 66
  non-specific medium effects on, 76
  paramagnetic contributions to, 49, 60, 64, 66
  specific medium effects on, 76
  theoretical background to, 47
Nuclear spin angular momentum, 2
Nuclear spin relaxation, 7
Nuclear spin–spin coupling, 6, 43
  contact, orbital and dipolar contributions to, 44, 85, 95, 103
  theoretical background to, 83

Orbital contribution to nuclear spin–spin coupling, 44, 83, 85, 95
Overlap integral, 30

Para-condensed benzenoid hydrocarbons,
  proton shielding of, 77
Paramagnetic contribution to nuclear shielding, 49
  one-centre contribution to, 60
  three-centre contribution to, 60
  two-centre contribution to, 60
Paramagnetic relaxation interactions, 12
Pauli exclusion principle, 28, 39
Pauling's island method, 40
Perturbation methods, 24
Phosphine, ^{31}P nuclear shielding of, 57
Phospholipids, ^{31}P nuclear shielding study of, 60
Platinum compounds, nuclear shielding anisotropy of, 16
Pople's shielding model, 63

Pyridine, 4
  ^{15}N relaxation of, 4
  nitrogen shielding anisotropy of, 16

Quantum chemistry,
  formalism of, 21

Radio-frequency magnetic field, B_1, 12
Ramsey's shielding model, 47
  AEE applied to, 51
  gauge dependence of contributions to, 50
Rayleigh–Schrödinger perturbation theory, 48, 51
Relativistic effects on nuclear shielding, 80
Resonance condition, 2
Ring current shielding tables, 74
Ring current contributions to nuclear shielding, 59, 69, 73
  quantum mechanical theory of, 71
Roothaan procedure, 30
Rumer–Pauling schemes, 96

s-orbital densities at the nucleus,
  values of for free atoms, 93
Scalar coupling relaxation, 16
Schrödinger's equation, 21
Second-order molecular properties,
  applications of quantum chemistry to, 40
Second-order perturbation terms, 24
Second-rank coupling tensor, 44
Secular determinant, 24
Self-consistent perturbation (SCP) calculations, 102
  applications of to calculations of spin–spin couplings, 94
Semi-empirical molecular orbital methods, 33
Slater–Condon repulsion parameters, 35
Slater determinant, 28
Slater's rules, 67
Slater-type orbitals, 36
Sodium chloride,
  ^{35}Cl line width of, 14

Solvaton model, 78
  applied to nuclear shielding calculations, 78
  applied to spin–spin coupling calculations, 108
Specific solvent effects on nuclear shielding, 76
Spectral analysis, 3
Spin Hamiltonian operator, 4
Spin rotation relaxation, 15
Spin rotation tensor, 56
Spin–spin couplings,
  angular dependence of, 106
  contact contribution to, 44, 83, 84, 94, 96
  dipolar contribution to, 44, 83, 85, 95
  hybridization effects on, 104
  orbital contribution to, 44, 83, 85, 95
  through-space contribution to, 108
Sum over states (SOS) perturbation theory, 48, 51
  applied to nuclear shielding calculations, 51, 63
  applied to spin–spin coupling calculations, 88, 99
Supermolecule approach to specific solute–solvent interactions, 76

Taylor series, 25, 26, 48
Tetrachloroethane,
  spin–spin coupling calculations of, 110
Through-space contribution to spin–spin couplings, 108
Townes and Dailey's method for estimating electric field gradients, 14
Triethylphosphine,
  ^{31}P relaxation of, 15
Trifluoromethane,
  spin–spin coupling calculations of, 103
Trimethyllead chloride,
  lead relaxation of, 16
Two-centre Hamiltonian, 61

Unrestricted SCF method, 31

Valence bond (VB) method, 39
    applied to spin–spin coupling
        calculations, 96
Variation method, 22
Vector potential due to applied
    magnetic field, 41

Water,
    ab initio shielding calculations of, 55
    spin–spin coupling calculations of, 101
Wavefunction, 21, 28